P9-BYL-075

Also from Visible Ink Press

The Handy Science Answer Book, 2nd edition

Can any bird fly upside down? Is white gold really gold? Compiled from the ready-reference files of the Science and Technology Department of the Carnegie Library of Pittsburgh, this best seller answers 1,400 questions about the inner workings of the human body and outer space, about math and computers, and about planes, trains, and automobiles. By the Science and Technology Department of the Carnegie Library of Pittsburgh, 7.25" x 9.25", 598 pages, 100 illustrations, dozens of tables, $16.95, ISBN 0-7876-1013-5.

The Handy Weather Answer Book

What's the difference between sleet and freezing rain? Do mobile homes attract tornadoes? What exactly is wind chill and how is it figured out? How can the temperature be determined from the frequency of cricket chirps? You'll find clear-cut answers to these and more than 1,000 other frequently asked questions in *The Handy Weather Answer Book*. By Walter A. Lyons, 7.25" x 9.25", 398 pages, 75 illustrations, $16.95, ISBN 0-7876-1034-8.

The Handy Space Answer Book

Is there life on Mars? Did an asteroid cause the extinction of dinosaurs? Find answers to these and 1,200 other questions in *The Handy Space Answer Book*. It tackles hundreds of technical concepts—quasars, black holes, NASA missions, and the possibility of alien life—in everday language. With vivid photos, thorough indexing, and an appealing format, it's as fun to read as it is informative for space lovers and curious readers of all ages. By Phillis Engelbert and Diane L. Dupuis, 7.25" x 9.25", 590 pages, $17.95, ISBN 1-57859-017-5.

The Handy Bug Answer Book

For anyone who's asked the question, "Why is the world full of bugs?" and for the child in the rest of us, there's *The Handy Bug Answer Book*. Author Dr. Gilbert Waldbauer shares his enthusiasm about the intricate and barely visible world of insects in this entertaining and accessible question-and-answer book. Organized by highly browsable topics, *Handy Bug* answers nearly 800 questions on insect lives and habits, including their number, sex lives, physical makeup, where they can be found, which are pests and which are good to have around, and the differences between insects, spiders, millipedes, and other invertebrates. By Dr. Gilbert Waldbauer, 7.25" x 9.25", 308 pages, 70 black-and-white and color illustrations, $19.95, ISBN 1-57859-049-3.

The Handy Earth Answer Book

The Earth is the world's biggest celebrity. Natural disasters . . . weather . . . global warming . . . and man's curious relationship with the highest mountains, deepest oceans, and extremes of heat and cold are always in the news. To satisfy everyone's Earthly desires, there's *The Handy Earth Answer Book*. Filled with plain-language answers to 1,000 commonly asked questions, *Handy Earth* is an easy-to-use resource with broad appeal. Parents, students, and armchair scientists will enjoy this highly browsable reference, which covers physical geology, mountains, plate tectonics, geophysics, geochemistry, oceanography, and other subjects in straightforward, question-and-answer format. By Dr. John Ernissee and Dr. Frank Vento, 7.25" x 9.25", 400 pages, 100 black-and-white and color photos, $19.95, ISBN 1-57859-0493.

The Handy Geography Answer Book

As the most up-to-date geographic reference available, *The Handy Geography Answer Book* offers nontechnical explanations of the natural features of the world and man's ever-changing mark on the planet, with maps and information on nations and states. Skillfully combining entertainment and information, it covers common trivia questions—highest, tallest, deepest, hottest, etc.—but also explores the social and political landscape, ranging from the influence of geography on language, religion, and architecture to how migration, population, and locations of countries and cities has been affected by terrain. By Matt T. Rosenberg, 7.25" x 9.25", 450 pages, 110 black-and-white and color photos and maps, $19.95, ISBN 1-57859-062-0.

The Handy Sports Answer Book

What makes a curveball curve? What was the longest basketball winning streak of all time? Which Supreme Court justice once won the NFL rushing title? When did professional baseball begin? Despite an endless stream of encyclopedias and single-sport references, few, if any, speak to the most commonly asked questions surrounding America's favorite sports. *The Handy Sports Answer Book* fills that gap by providing an easy-to-use, entertaining, and broadly based reference that sports fans will use to expand their knowledge, extend their interests, settle arguments, and hone their trivia skills. By Kevin Hillstrom, Laurie Hillstrom, and Roger Matuz, 7.25" x 9.25", 446 pages, 150 black-and-white and color photos, $19.95, ISBN 1-57859-075-2.

THE HANDY PHYSICS ANSWER BOOK

THE HANDY PHYSICS ANSWER BOOK™

P. Erik Gundersen

The Handy Physics Answer Book™

Published by Visible Ink Press
42015 Ford Road #208
Canton, MI 48187-3669

Most Visible Ink Press books are available at special quantity discounts when purchased in bulk by corporations, organizations, or groups. Customized printings, special imprints, messages, and excerpts can be produced to meet your needs.

Art Director: Michelle DiMercurio
Typesetting: The Graphix Group

ISBN 1-57859-106-6

Printed in the United States of America

10 9 8 7 6 5 4 3 2

This book is dedicated to the memory of Phil Goodyear (1941–1997),
an extraordinary teacher, colleague, and mentor,
whose enthusiasm for life and teaching physics
has made an everlasting impression on me.

Contents

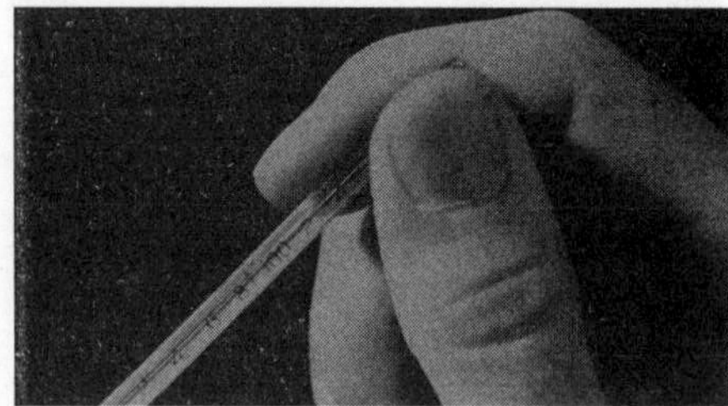

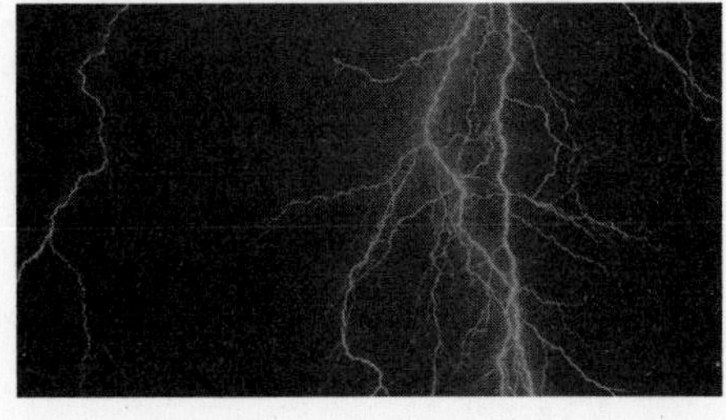

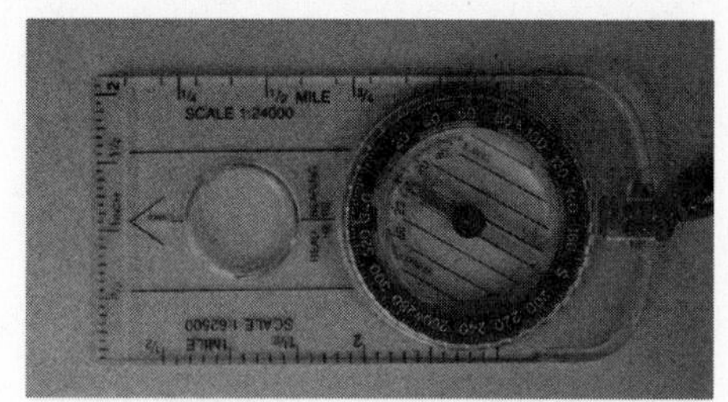

1/2 MILE
SCALE 1:24000
SCALE 1:62500

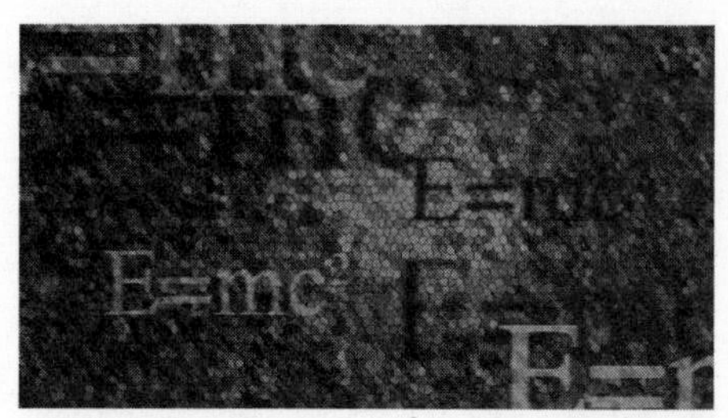

E=mc²

Introduction

Physicists—the really good ones, like Albert Einstein—are known for asking simple questions. One of Einstein's first thoughts on light came at the age of five. He asked, "I wonder what the world would look like while riding on a wave of light?" Einstein spent the rest of his life asking and seeking the answers to the most basic questions regarding how the universe works. Sheldon Glashow, a Nobel Prize–winning physicist and professor at Harvard, once said that physicists are like children. Children have a tendency to wonder and therefore ask many questions that adults might find too fundamental. Since the art and science of physics involve the quest for the underlying or fundamental principles of the universe, a characteristic of intense questioning is a handy one for physicists. With that in mind, *The Handy Physics Answer Book* examines more than 825 basic questions about physics and physicists, ranging from the application of physics in everyday life to the latest explorations in subatomic physics.

Why would it be dangerous to drop a penny off the Empire State Building? (p. 39). Do you know what causes a curve ball to curve? (p. 140). How do ice skates work? (p. 40). Where are the largest tide fluctuations? (p. 50). How does an object stay in orbit around the Earth? (p. 63). How fast would a baseball have to be hit in order for it to be put into orbit? (p. 64). What is fluid dynamics? (p. 136). What is a shock wave? (p. 144). What is the lowest possible temperature? (p. 157). What is the relationship between frequency, wavelength, and velocity? (p. 175). How do 3-D movies work? (p. 249). Why is a car often the best place to be when lightning strikes? (p. 299). Hint: It's not the rubber tires.

The Handy Physics Answer Book moves away from the mathematical explanations often associated with physics and instead takes a more conceptual approach—written in everyday English. The book begins with a **General** section dealing with the basics of physics, such as "What is physics?" and "What does a physicist do?," and then explores the contributions of some of the legendary physicists in a series of questions dealing with the coveted Nobel prize.

Handy Physics then turns to **Movement**, where the topics of speed, gravity, and momentum are addressed in such questions as, "How does the moon affect the tides?"

and "Why do airbags save lives?" The chapter **Work, Energy, and Simple Machines** follows. **Objects at Rest** answers questions such as "Why are football players and wrestlers taught to get down low when blocking or making a move on someone?" and "What is the newest hybrid of bridges?." "Why are microbursts so dangerous to airplanes?" is a typical question from the **Fluids** chapter. **Heat and Thermodynamics** addresses such difficult topics as "Why do water droplets accumulate on the outside of your water glass?" and "How do refrigerators keep food cool?"

In the subsequent chapters **Waves, Sound,** and **Light,** we shift gears to answer questions dealing with such topics as wave motion, how we hear, and how rainbows are formed. **Electricity** tackles questions concerning electrical shocks and circuits, while **Magnetism, Electromagnetism, and Electronics,** discusses magnetic levitation, metal detectors, and how a compass works.

The "cutting edge" of scientific discoveries and breakthroughs is discussed in **Modern Physics.** Everything from a quanta to nuclear reactions to the latest discoveries of subatomic particles are addressed in this section. The final chapter of *Handy Physics* specifically concerns itself with brilliant and often bizarre theories that physicists such as Albert Einstein and Stephen Hawking have developed. Questions addressed in this **Deep Theories** chapter include "What important breakthrough has been made in the observation of neutrinos?," "What do physicists believe will eventually happen to the universe?," and various questions dealing with Einstein's ideas about time travel.

As a physics teacher at Pascack Valley High School in Hillsdale, New Jersey, I know how important it is to make physics fun, exciting, and relevant to my students' lives. I have attempted to do the same with this *Handy Answer* book. Whether reading the book straight through or simply flipping through the pages, these questions and answers will help you think about the world "the physics way." Maybe you'll be walking down the street one day and start seeing all the physics around you. You might see a car driving down the road and wonder, "Is the car moving, am I moving, or are we both moving relative to each other?" Spotting one of those nasty cumulonimbus thunderhead clouds looming overhead, you'll think, "I better get inside a car since it acts like a Faraday cage." Or perhaps on a sunny day you might just feel it necessary to explain to someone why the sky is blue, why the clouds are white, and why the colors of the rainbow are always in the same order. For the physics fan, this book contains much knowledge and trivia.

Always remember, as physicists say, "Physics is Phun!"

Acknowledgments

I wrote *The Handy Physics Answer Book* while teaching high school physics full time. I am grateful to many people who helped me with the writing, editing, production, and marketing of this book, as well as to those who lent their support and enthusiasm.

Someone who can testify to how much time was spent on writing and teaching this past year is my wife Amy. Amy's encouragement, patience, and love, not to mention her suggestions for questions, proofreading abilities, constructive criticism, and compliments, proved to be a huge help while writing this book. I could not have done this without her.

A number of other people have assisted with the development and writing of *The Handy Physics Answer Book*. Appreciation needs to be extended to Al Cann, David Tedesco, and the rest of the Pascack Valley High School faculty and my 1997–98 physics students for their suggestions, support, and enthusiasm while writing this book.

A great deal of thanks also needs to be extended to Peder and Kirsten Gundersen for their encouragement and proofreading, as well as to Gary and Barbara Frye for reviewing sections of the book to make sure it made sense. Thank you Frank Mortimer for entertaining my questions about the publishing world and assisting me with some of the details of the book.

Much gratitude is extended to Visible Ink Press senior editor Carol Schwartz, who provided a huge amount of support and assistance throughout the writing of this book. Thanks to Betsy Rovegno and Julia Furtaw for allowing me the opportunity to write this book. And thanks to other VIP staff members who helped significantly with the development of this *Handy* book: Matt Nowinski for proofing and fact checking, Edna Hedblad for photo research and permissions, Pam Reed and Randy Bassett for photo processing, art director Michelle DiMercurio, and Jeff Muhr for technical support. Also thanks to Kim Marich, Lauri Taylor, and Marilou Carlin for their work in getting this book to you, the reader.

Copy editor Don Roberts, with his incredible knowledge of physics and his suggestions, compliments, and constructive criticism, was an invaluable resource in the writing of *The Handy Physics Answer Book.* His revisions of questions and answers, especially in the field of aviation, modern physics, and thermodynamics, helped significantly and will contribute to the success of this book.

Finishing touches on the book were provided by typesetter Marco Di Vita, indexer Barbara Cohen, proofreader Brigham Narins, and photographer Jim Olenski.

This acknowledgment would not be complete without thanking those who showed me what good physics teaching is all about. If it were not for Professor Barnhill at the University of Delaware, Pete Parlett of McKean High School in Wilmington, Delaware, and Phil Goodyear of Pascack Valley High School, I would not be teaching physics and would certainly not have been able to write *The Handy Physics Answer Book.*

Photo Credits

Front cover photos: Albert Einstein, reproduced with permission from UPI/Corbis-Bettmann; prism, photograph by Peter Angelo Simon, reproduced with permission from Stock Market; electrical current, © Phil Jude, National Audubon Society Collection/Photo Researchers, Inc., reproduced with permission.

Acceleration of projected and dropped balls, strobe photograph by Richard Megna. © 1990 Richard Megna, Fundamental Photographs. Reproduced with permission.

Air bag deploying from passenger side, photograph. AP/Wide World Photos. Reproduced with permission.

Amusement park ride, photograph. © Eunice Harris. Photo Researchers, Inc. Reproduced with permission.

Arecibo Observatory, photograph. UPI/Corbis-Bettmann. Reproduced with permission.

Aristotle, photograph of a woodcut. UPI/Corbis-Bettmann. Reproduced with permission.

Atomic bombs Fat Man and Little Boy, photograph. Courtesy the Library of Congress.

Atomic clock, photograph. UPI/Corbis-Bettmann. Reproduced with permission.

Banked roadway at Daytona International Speedway, photograph. Archive Photos, Inc. Reproduced with permission.

Bay of Fundy at low tide, photograph. Archive Photos, Inc. Reproduced with permission.

Bed of nails, photograph. Archive Photos, Inc. Reproduced with permission.

Big bang, illustration by Ludek Pesek. National Audubon Society Collection/Photo Researchers, Inc. Reproduced with permission.

Big bang theory, illustration. AP/Wide World Photos. Reproduced with permission.

Black hole (core of Whirlpool Galaxy M51), photograph. Courtesy U.S. National Aeronautics and Space Administration (NASA).

Cat falling, stroboscopic photograph. © J. P. Varin/Jacana. Photo Researchers, Inc. Reproduced with permission.

Catcher's mitt: Al Lopez, Cleveland Indian's catcher, photograph. AP/Wide World Photos. Reproduced with permission.

Chernobyl: demonstrator wearing skeleton costume, Frankfurt, Germany, photograph. AP/Wide World Photos. Reproduced with permission.

Circular pulley (block and tackle), photograph by Grant Smith. UPI/Corbis-Bettmann. Reproduced with permission.

Color-coded resistors, photograph. © Jim Olenski. Reproduced with permission.

Compass, photograph. © Robert J. Huffman/Field Mark Publications. Reproduced with permission.

Condensation forming on glass of water, photograph. © Jim Olenski. Reproduced with permission.

Copernicus' view of the Earth's revolutions around the sun, illustration. UPI/Corbis-Bettmann. Reproduced with permission.

Cornell, Chris (of Soundgarden), photograph by Ken Settle. © Ken Settle. Reproduced with permission.

Curie, Marie, with daughter Irène Joliot-Curie, photograph. UPI/Corbis-Bettmann. Reproduced with permission.

Dame Point Bridge, photograph. © Tom Carroll/Phototake NYC. Reproduced with permission.

Detail of human eye, diagram by Hans & Cassidy. Gale Research.

Doppler effect, illustration. Gale Research.

Eclipse as viewed through pin-hole camera, photograph. © Robert J. Huffman/Field Mark Publications. Reproduced with permission.

Eclipse, lunar, five phases over Toronto, Canada, photograph. Reuters/UPI/Corbis-Bettmann. Reproduced with permission.

Eclipse, solar, multiple exposures, photograph. UPI/Corbis-Bettmann. Reproduced with permission.

Eclipse, total solar, photograph. Courtesy New York Public Library Picture Collection.

Edison, Thomas, in West Orange, New Jersey, laboratory, photograph. UPI/Corbis-Bettmann. Reproduced with permission.

Einstein, Albert, photograph. Courtesy the Library of Congress.

Einstein, Albert, and David Ben-Gurion, photograph. UPI/Corbis-Bettmann. Reproduced with permission.

Electric chair at Somers, Connecticut, photograph. UPI/Corbis-Bettmann. Reproduced with permission.

Electric eel, photograph. UPI/Corbis-Bettmann. Reproduced with permission.

Electromagnetic radiation spectrum, illustration by Robert L. Wolke. Reproduced with permission.

Electromagnetic wave, San Francisco Exploratorium, photograph. Phil Schermeister/UPI/Corbis-Bettman. Reproduced with permission.

Fermi, Enrico, photograph. Courtesy the Library of Congress.

Firth of Forth Cantilever Bridge, Edinburgh, Scotland, photograph. UPI/Corbis-Bettmann. Reproduced with permission.

Galileo, portrait based on painting by U. Susterman, photograph. UPI/Corbis-Bettmann. Reproduced with permission.

Galileo's telescope, reconstruction, photograph. Jim Sugar Photography/UPI/Corbis-Bettman. Reproduced with permission.

Gears in a film projector, photograph. © Jim Olenski. Reproduced with permission.

Geiger counter: man taking readings from Susquenna River near Three Mile Island nuclear power plant, photograph. UPI/Corbis-Bettmann. Reproduced with permission.

Golf ball, photograph. © Jim Olenski. Reproduced with permission.

Goodyear blimp, airship "Spirit of Akron," photograph. UPI/Corbis-Bettmann. Reproduced with permission.

Great horned owl, photograph. David A. Northcott/UPI/Corbis-Bettman. Reproduced with permission.

Ground fault interrupt (GFI) outlet, photograph. © Jim Olenski. Reproduced with permission.

Gyroscope, photograph. Ed Carlin/Archive Photos, Inc. Reproduced with permission.

Hawking, Stephen, photograph. AP/Wide World Photos. Reproduced with permission.

Hearing, human system, diagram by Hans & Cassidy. Gale Research.

Hindenburg explosion, photograph. AP/Wide World Photos. Reproduced with permission.

Hologram, photograph. © Robert J. Huffman/Field Mark Publications. Reproduced with permission.

Hubble Space Telescope, photograph. © 1996 UPI/Corbis-Bettman. Reproduced with permission.

Hydraulic lift used by farmer, photograph. AP/Wide World Photos. Reproduced with permission.

Igloo, photograph. UPI/Corbis-Bettmann. Reproduced with permission.

Iron cross: Vitaly Scherbo performing gymnastic routine at Barcelona Summer Olympics, July 31, 1992, photograph. AP/Wide World Photos. Reproduced with permission.

Laser surgery, photograph. Kevin Fleming/UPI/Corbis-Bettman. Reproduced with permission.

Leaning Tower of Pisa, photograph. AP/Wide World Photos. Reproduced with permission.

Lever and fulcrum, illustration. UPI/Corbis-Bettmann. Reproduced with permission.

Leyden jar engraving, photograph. UPI/Corbis-Bettmann. Reproduced with permission.

Lightning strikes on Tucson horizon, photograph. Photo Researchers, Inc. Reproduced with permission.

Marconi, Guglielmo, photograph. Courtesy the Library of Congress.

McCandless, Bruce, participating in first use of manned maneuvering unit, photograph. Courtesy U.S. National Aeronautics and Space Administration (NASA).

Mercury droplets, photograph. © Dr. Jeremy Burgess/Science Photo Library, National Audubon Society Collection/Photo Researchers, Inc. Reproduced with permission.

Michelson, A. A., with Albert Einstein, R. A. Millikan, (back row) Dr. W. S. Adams, W. Mayer, Dr. M. Farrand (1942), photograph. Archive Photos, Inc. Reproduced with permission.

Microwave oven, diagram by Hans & Cassidy. Gale Research.

Microwave oven, protective screen on door, photograph. © Jim Olenski. Reproduced with permission.

Mirage in desert, photograph. O. Alamany & E. Vicens/UPI/Corbis-Bettman. Reproduced with permission.

Monkey, photograph. AP/Wide World Photos. Reproduced with permission.

Movers using ramp as an inclined plane, photograph. UPI/Corbis-Bettmann. Reproduced with permission.

Music CDs, photograph. © Jim Olenski. Reproduced with permission.

Newton, Sir Isaac (oval frame, allegorical elements surrounding), engraving. Courtesy the Library of Congress.

Newton, Sir Isaac, illustration. © Mikki Rain. Photo Researchers, Inc. Reproduced with permission.

One-way mirror, photograph. © Spencer Grant. Photo Researchers, Inc. Reproduced with permission.

Optical fibers, photograph. Roger Ressmeyer/UPI/Corbis-Bettman. Reproduced with permission.

Pont du Gard, photograph. UPI/Corbis-Bettmann. Reproduced with permission.

Prism: light shining through glass prism, forming a rainbow, diagram by Hans & Cassidy. Gale Research.

Racing tires: mechanic pushing slick tires at Twin Ring Motegi Superspeedway, March 27, 1998, photograph. AP/Wide World Photos. Reproduced with permission.

Refraction of swizzle stick in water, photograph. © Jim Olenski. Reproduced with permission.

Roller coaster, Wildwood, New Jersey, photograph. Archive Photos, Inc. Reproduced with permission.

Rotating space station/farming spaceship, illustration. Julian Baum/Science Photo Library/Photo Researchers, Inc. Reproduced with permission.

Rotating space station in orbit, illustration. David Hardy/Science Photo Library/Photo Researchers, Inc. Reproduced with permission.

Rotation around a center of mass: a spinning wrench, photograph by Richard Megna. © 1990 Richard Megna, Fundamental Photographs. Reproduced with permission.

Screws, photograph. © Jim Olenski. Reproduced with permission.

Shanghai World Financial Centre, China (artist's rendering), illustration. Reuters/HO/Archive Photos, Inc. Reproduced with permission.

Skater: Michelle Kwan at Winter Olympics in Nagano, Japan, February 18, 1998, photograph. AP/Wide World Photos. Reproduced with permission.

Skydivers before parachutes open, photograph. Express Newspapers/Archive Photos, Inc. Reproduced with permission.

Solar energy panels at the National Renewable Energy Laboratory in Golden, Colorado, photograph. U.S. Department of Energy, Washington, D.C.

Sphygmomanometer: man getting his blood pressure tested, photograph. UPI/Corbis-Bettmann. Reproduced with permission.

Stealth bomber, photograph. U.S. National Aeronautics and Space Administration (NASA).

Streamline testing of Ferrari F300 in a windtunnel, photograph. AP/Wide World Photos. Reproduced with permission.

Sundial, photograph. UPI/Corbis-Bettmann. Reproduced with permission.

Superconductor, photograph. © Takoshi Takahara. Photo Researchers, Inc. Reproduced with permission.

Supernumerary rainbow over Calanche Cliffs, photograph. Kevin Schafer/UPI/Corbis-Bettman. Reproduced with permission.

Surfer: Steve McDonald under tubular wave, photograph. Tony Arruza/UPI/Corbis-Bettman. Reproduced with permission.

Tacoma Narrows Bridge, just before collapse, photograph. UPI/Corbis-Bettmann. Reproduced with permission.

Thermal expansion, iron and aluminum, illustration by Hans & Cassidy. Gale Research.

Thermometer, photograph. Archive Photos, Inc. Reproduced with permission.

Three Mile Island's Edison Nuclear Power Plant, photograph. AP/Wide World Photos. Reproduced with permission.

Tokamak fusion test reactor vacuum vessel, photograph. AP/Wide World Photos. Reproduced with permission.

Treadmill, illustration from 17th century, photograph. UPI/Corbis-Bettmann. Reproduced with permission.

Tree swallows on electrical wires, photograph. © Robert J. Huffman/Field Mark Publications. Reproduced with permission.

Ultrasound of fetus. Courtesy of Jacqueline Longe.

Van Allen Belts (magnetic lines of force), diagram. Gale Research.

Van de Graaff generator, photograph. Lester V. Bergman. UPI/Corbis-Bettman. Reproduced with permission.

Verrazano-Narrows Bridge, photograph. UPI/Corbis-Bettmann. Reproduced with permission.

Water tower in West Branch, Michigan, photograph. AP/Wide World Photos. Reproduced with permission.

Windmills, photograph by James Perez. Greenpeace. Reproduced with permission.

Wright, Wilbur, and Orville Wright (in their airplane), photograph. Courtesy the Library of Congress.

Yeager, Chuck, standing next to Bell X-1 airplane, photograph. UPI/Corbis-Bettmann. Reproduced with permission.

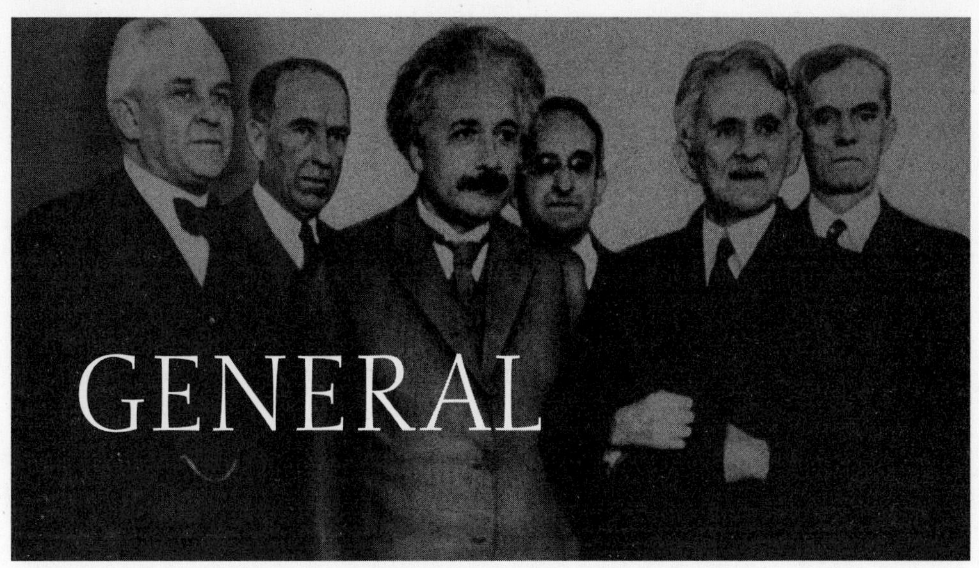

THE BASICS

What is **physics**?

Physics is often considered the basis for all sciences. It studies and describes the motion, energy, momentum, and forces on matter throughout the entire universe. Many scientists believe that in order to truly understand the other sciences (biology, chemistry, geology, astronomy, etc.) one must first have an understanding of physics. For example, in biology, the movement of blood is related to motion, gravity, and fluid dynamics—all areas of study in physics. In astronomy, the motion of the planets, stars, and galaxies all rely on the Law of Universal Gravitation. Physics has a place in all the sciences, which explains why physics is often regarded as the fundamental science.

What are the **subdivisions** of physics?

It was not until the 1800s that physics was recognized as a science all its own. Before the nineteenth century, a physicist was called a "natural philosopher" and worked in other fields such as mathematics, philosophy, biology, or chemistry. However, since the early 1800s, physics has separated from the other sciences and has proven to be a very large and important field of study. Due to the huge breadth of physics, approximately seventeen divisions in physics have emerged.

Mechanics	The primary field of physics. Mechanics deals with the cause and effect of forces, motion, and energy on physical objects.
Thermodynamics	"Thermo," as it is called for short, concentrates on the study of heat and how heat energy is transformed from one form of energy to another.
Cryogenics	The study of matter at extremely low temperatures.
Plasma Physics	Examines the activity of highly ionized, electrically charged gases.
Solid State Physics	Otherwise known as condensed matter physics, solid state physics is the study of the physical properties of solid materials.
Geophysics	Geophysics is the physics of the Earth. It deals with the forces and energy found within the Earth itself and how they impact the Earth. Geophysicists study earthquakes, volcanic activity, and oceanography.
Astrophysics	Studies how interstellar bodies, such as planets and stars, interact with one another.
Acoustics	The study of sound and how sound travels.
Optics	The study of light and how light travels.
Electromagnetism	The study of the interaction between electric and magnetic fields, and the electric charges that produce those fields.
Fluid Dynamics	Observes the behavior of moving liquids and gases.
Mathematical Physics	Relates mathematical processes to physics.
Statistical Mechanics	Statistics are used to model the effects of systems that are composed of many particles.
High Energy Physics	Dedicated to searching for fundamental particles.
Atomic Physics	Uses the knowledge of the fundamental particles to understand the structure of the individual atom.
Molecular Physics	Takes the knowledge of atomic physics and applies it to understand the structure of molecules.
Nuclear Physics	The investigation into the structure of the atomic nucleus, nuclear reactions, and their applications.
Quantum Physics	The study of extremely small systems and the quantization of energy.

What is the difference between science and technology?

Often people confuse science and technology. Science is a field in which information is gathered and hypotheses are formulated through the experimentation, observation, and cataloging of information and ideas. Technology is a field that strives to satisfy human material needs or desires. It is from science that technology gets the information it needs to satisfy our need for "progress." Were it not for science, there would be no technology, and many people feel that it is the desire for increased technology that drives science.

MEASUREMENT

What are the **standards for measurement** in physics?

The International System of Units, officially known as Système International and abbreviated as SI, was adopted by the eleventh General Conference on Weights and Measures in Paris, 1960. Basic units were to be based on the meter-kilogram-second (MKS) system. This is commonly known as the metric system.

Why doesn't it seem as though the **United States** uses the **SI system**?

Although the American scientific community uses the SI system of measurement, the general American public still uses the traditional English system of measurement. In an effort to change over to the metric system, the United States government instituted the Metric Conversion Act in 1975. Although the act committed the United States to increasing the use of the metric system, it was on a voluntary basis only. The Omnibus Trade and Competitiveness Act of 1988 required all federal agencies to

How is the length of a meter determined?

Today the meter is the standard for both the International System of Units and the metric system. The meter was originally defined as the distance between two lines on a bar of platinum-iridium alloy, 1/10,000,000th the distance from the North Pole to the equator. However, the bar was deemed inaccurate because it would expand and contract depending on its temperature.

In 1960, scientists determined that light would be a better method of measuring a meter. As a result, the meter was defined as the distance of 1,650,763.73 wavelengths of reddish-orange light emitted by the krypton-86 isotope. The standard for measuring the meter was changed again in 1983 as the distance light travels in a vacuum in 1/299,792,458th of a second.

adopt the metric system in their business dealings by 1992. Therefore, all companies that held government contracts had to convert to metric. Although approximately 60% of American corporations manufacture metric products, the English system still seems to be the predominant system of measurement in the United States.

Who defined or developed **the meter**?

In 1798, French scientists determined that the meter would be measured as 1/10,000,000th the distance from the North Pole to the Equator. After calculating this distance, the scientists made a platinum-iridium bar that measured that exact distance. This standard was used until 1960, when newer and more accurate methods of measuring the meter evolved.

What is the standard unit for measuring **mass**?

The kilogram is the standard unit for mass in the metric system and the International System of Units. The kilogram was originally defined as

An atomic clock (see next page).

the mass of 1 cubic decimeter of pure water at 4° Celsius. A platinum cylinder, the same mass as the cubic decimeter of water, was the standard until 1889 when the platinum cylinder was replaced by a platinum-iridium cylinder. This cylinder, whose mass is close to the original platinum cylinder, is permanently kept near Paris.

How is a **second** measured?

Atomic clocks are the most precise devices to measure time. Atomic clocks such as rubidium, hydrogen, ammonia, and cesium clocks are used by scientists and engineers when computing distances with Global Positioning Systems (GPS) and measuring the rotation of the Earth.

The most stable measurement clock that is used as the standard for the second is the cesium-133 atomic clock. The measurement of the second is defined as the time it takes for 9,192,631,770 periods of microwave radiation to transfer the cesium-133 atom from a lower-energy state to a higher-energy state.

What was the **first clock**?

The first method of measuring time dates back to 3500 B.C., when a device known as the gnomon was used. The gnomon was a stick placed vertically into the ground which, when struck by the sun's light, produced a distinct shadow. By measuring the relative positions of the shadow throughout the day, the time of day was able to be monitored. The gnomon was later replaced by the first hemispherical sundial in the third century B.C. by the astronomer Berossus.

The problem with gnomons and sundials became apparent when the sun disappeared. To remedy this problem, timing devices such as notched candles were created. Later, hourglasses and water clocks (clepsydra) became quite popular.

Also in the third century B.C. , Ctesibius of Alexandria, a Greek inventor, created a primitive mechanical clock that used gears to keep accurate time.

A sundial.

What do some of the **metric prefixes** represent?

Prefixes in the metric system are used to denote powers of ten. The value of the exponent next to the number ten represents the number of places the decimal should be moved to the right (if the number is positive), or to the left (if the number is negative). The following is a list of prefixes commonly used in the metric system:

pico	(p)	10^{-12}	deka	(da)	10^{1}
nano	(n)	10^{-9}	hecto	(h)	10^{2}
micro	(μ)	10^{-6}	kilo	(k)	10^{3}
milli	(m)	10^{-3}	mega	(M)	10^{6}
centi	(c)	10^{-2}	giga	(G)	10^{9}
deci	(d)	10^{-1}	tera	(T)	10^{12}

How does **"accuracy"** differ from **"precision"**?

Both "accuracy" and "precision" are often used interchangeably in everyday conversation; however, each has a unique meaning. Accuracy defines how correct or how close to the accepted result or standard a measurement or calculation has been. Precision deals with how easily the results can be reproduced. For example, a person who can hit a bull's eye with a bow and arrow is accurate. However, if the person cannot repeatedly hit the bull's eye, then s/he is not precise.

PHYSICISTS

Careers in Physics

How does one **become a physicist**?

The first requirement to be a physicist is having an inquisitive mind. Albert Einstein himself admitted, "I'm like a child. I always ask the simplest questions." It seems as though the simplest questions always seem to be the most difficult to answer.

These days, becoming a physicists requires quite a bit of schooling along with that inquisitive mind. In high school, a strong academic background including mathematics, English, and science is necessary in order to enter college with a strong knowledge base. Once in college as a physics major, courses such as classical mechanics, electricity and magnetism, optics, thermodynamics, modern physics, and calculus are required for a bachelor's degree.

To become a research physicist, an advanced degree is usually required. This means attending graduate school, performing research, writing a thesis, and eventually obtaining a Ph.D. (Doctor of Philosophy).

What does one **do as a physicist**?

Physicists can find employment in a variety of fields. Many research physicists work in environments where they perform basic research. These scientists typically work in industry, research universities, and astronomical observatories. Physicists who find new ways to use physics are often employed by engineering, business, law, and consulting firms.

What is the difference between theoretical physicists and experimental physicists?

Physicists who make predictions based on previous laws and theories are called theoretical physicists. However, the physicist who attempts to verify (and extend or revise) theories through experimentation is called an experimentalist. For example, Albert Einstein was considered probably the greatest theoretical physicist of all time, but he was not well known for his experiments in the laboratory. On the other hand, another great physicist, Galileo Galilei, consistently verified his own theories in the laboratory.

Physicists are also extremely valuable in areas such as computer science, medicine, communications, and publishing. Finally, many physicists who love to see young people get excited about physics become teachers.

What jobs do **non-physicists** hold that use physics every day?

Every job has some relation to physics, but there are some examples that many would not think of as being physics-intensive. Athletes, both professional and amateur, use the principles of physics all the time. The laws of motion to lift, throw, push, hit, tackle, run, drag, jump, and crawl are present all the time. The more an athlete and coach understands and uses their knowledge of physics in their sport, the better that athlete will become.

Automotive mechanics use physics concepts every day as well. In fact, areas such as optics, electricity and magnetism, thermodynamics, and mechanics play greater roles as vehicles become more and more complex.

Another field where physics is extremely important is in X-ray, CAT scan, computed tomography, and magnetic resonance imaging (MRI). The technicians in hospitals and medical clinics who use such technology must have an understanding of what X-rays and magnetic resonance imaging are, how they behave, and how such "high-tech" instruments are to be used.

Who was the first person to declare that the Earth was round?

Contrary to what many people think, Christopher Columbus was not the first person to declare that the Earth was round. Almost 1800 years before Columbus, Aristarchus of Samos not only declared the Earth round, but also suggested that it orbited the sun. In order to make such hypotheses, Aristarchus measured the angles between the sun and the surface of the Earth in two different cities and found that the angles differed dramatically. From this measurement and other calculations, Aristarchus became the first person to prove that the Earth was round.

FAMOUS PHYSICISTS

Who were the **first physicists**?

Although physics was not considered a distinct field of science until the early nineteenth century, people have been studying the motion, energy, and forces that are at play in the universe for thousands of years. The earliest documented accounts of serious thought toward physics, specifically the motion of the planets, dates back to the years of the Egyptians, Mesoamericans, and the Babylonians. The Greek philosophers Plato and Aristotle analyzed the motion of objects, but were not interested in performing experiments to prove or disprove their theories.

What contributions did **Aristotle** make?

Aristotle was a Greek philosopher and scientist who lived for sixty-two years in the fourth century B.C. He was a student of Plato and an accomplished scholar in the fields of biology, physics, mathematics, philosophy, astronomy, politics, religion, and education. In physics, Aristotle is famous for his belief that motion occurs because a body feels the need or

Aristotle.

desire to move, and that to change motion requires some sort of violent outside "cause." These ideas, although not completely correct, proved to be the beginnings of the study of motion that Galileo and Newton would refine hundreds of years later.

It was not until the third century B.C. and later that experimental achievements in physics were made in such cities as Alexandria and other major cities throughout the Mediterranean. Archimedes measured the density of objects by measuring their displacement of water. Aristarchus of Samos is credited with measuring the ratio of the distances from the Earth to the sun and to the moon. Erathosthenes determined the circumference of the

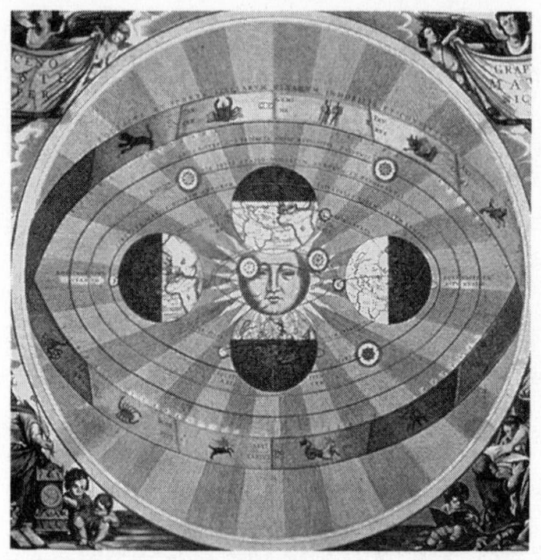

Copernicus' view of the Earth's revolutions around the sun.

Earth by using shadows and trigonometry. Hipparchus discovered the precession of the equinoxes. And finally, Ptolemy proposed an order of planetary motion in which the sun, stars, and moon revolved around the Earth.

How did Nicolas **Copernicus** view the solar system?

Nicolas Copernicus was the first person to view the solar system as a heliocentric (sun-centered) system instead of a geocentric (Earth-centered) system. In 1543, the same year as his death, he published *The Revolution of Heavenly Orbs.* His book was dedicated to the Pope, which was ironic since Copernicus' ideas were not supported by the Catholic Church. Surprisingly, Copernicus' book was not immediately banned by the church, probably because of a disclaimer in the book that Copernicus was not aware of. It said that the ideas in the book were only meant to make the calculations of the motion of the planets easier and in no way should be taken as what really occurs. Although Copernicus' ideas were indeed meant to be taken literally, the disclaimer did keep the book in circulation longer than expected.

What famous scientist was placed under **house arrest** for agreeing with Copernicus?

Galileo Galilei was responsible for bringing the Copernican system more recognition. In 1632, Galileo published his book *Dialogue Concerning the Two Chief World Systems.* The book originally had the approval of Catholic censors, but was banned by the Pope for siding with the Copernican model of the solar system. The Pope, a long-time friend of Galileo, tried Galileo and placed him under house arrest for the rest of his life.

Galileo was also famous for his work on motion; he is probably best known for his experiment on the Leaning Tower of Pisa, where he found that a heavy rock and a light rock fell at the same rate. This was revolu-

Galileo Galilei.

Sir Isaac Newton.

tionary thinking and began what many in the scientific world feel was the beginning of true physics.

Who is considered one of the **most influential scientists** of all time?

Many scientists and historians consider Isaac Newton one of the most influential people of all time. It was Newton who discovered the laws of motion and Universal Gravitation, made huge breakthroughs in light and optics, built the first reflecting telescope, and developed calculus. His discoveries published in *The Principia* and in *Optiks* are unparalleled and formed the basis for all physics until Einstein's theories of Relativity were developed in the early twentieth century.

Why did Newton **leave his studies** at Cambridge in 1661?

Newton was encouraged by his mother to become a farmer, but his uncle saw the talent Newton had for science and math and helped him enroll in Trinity College in Cambridge. Newton spent only two years there, however; he returned to his home town of Woolsthorpe to flee the spread of the Black Plague. It was in Woolsthorpe that Newton would make his most important discoveries.

What **official titles** did Newton receive?

Newton was extremely well respected in his time. Although he was known for being nasty and rude to his contemporaries, Newton became Lucasian Professor of Mathematics at Cambridge in the late 1660s, president of the Royal Society of London in 1703, and the first scientist ever knighted, in 1705.

Who would become the **most influential scientist** of the twentieth century?

On March 14, 1879, Albert Einstein was born in Ulm, Germany. No one knew that this little boy would one day grow up and change the way people viewed the laws of the universe. As a child, little Albert hated the regimented style of school. He would cut class and study the natural laws of physics and mathematics on his own. Since he never really impressed his teachers, Einstein was not awarded a university position after his days at the Polytechnic School in Zurich. Instead, Einstein became a patent clerk in Bern, Switzerland.

What did **Einstein** do to win such fame?

Einstein was a wonderful theoretical physicist. Among his accomplishments were the Special and General Theories of Relativity, in which he changed the assumptions upon which physics was based up until his time, creating a radically different set of physical laws. With his theories of Relativity, Einstein described how time can slow down and how mass is equivalent to energy. Although these were his most impressive theories, Einstein would win the Nobel prize in physics in 1921 as a result of his work on the photoelectric effect.

Why did Einstein win a **Nobel prize** for the photoelectric effect, but not for Relativity?

The rules of the Nobel prize at that time said that the award had to be for experimental work, which precluded Einstein's theoretical work in Relativity. It is believed that the Nobel committee rationalized Einstein's work with the photoelectric effect (which also was, for the most part, theoretical work—his true involvement with the experimental work was minimal) in order to award him a most-deserved Nobel prize.

Why was Einstein **more** than just a world-renowned **physicist**?

Einstein knew that his theories had significant consequences on the world. Einstein spoke out against his fellow Germans in World War I;

Albert Einstein and Israeli prime minister David Ben-Gurion.

and when Hitler came to power, Einstein decided to head for the United States. He became a citizen and took a position at the Institute for Advanced Study in Princeton, New Jersey. After collaborating with other physicists, Einstein authored a letter to President Roosevelt urging the United States to respond to Germany's desire to build an atomic bomb based on his formula $E = mc^2$ by building a bomb of their own.

Although Einstein did not actually work on the bomb, he regretted and felt partly responsible for the death and destruction in Japan. He sent another a letter to the President urging him not to use the bomb, but the letter was never read. After the war Einstein spent time lobbying for atomic disarmament. At one point he was even asked to head the new Jewish state of Israel. Einstein, both for his scientific works and his social and political views, became a national and international icon.

THE NOBEL PRIZE

What is the **Nobel prize**?

The Nobel prize is one of the most prestigious awards in the world. It was named after Alfred B. Nobel, the inventor of dynamite; he left $9,000,000 in trust, of which the interest was to be awarded to the person who made the most significant contribution to their particular field that year. The awards, given in the fields of physics, chemistry, physiology and medicine, literature, peace, and economics, comprise approximately $1,000,000, as well as a great deal of recognition.

Who were the **Nobel prize winners for physics in 1997**?

The winners of the Nobel prize in physics for 1997 were Steven Chu and William D. Phillips of the United States and Claude Cohen-Tannoudji of France. The three men shared the prize for their development in cooling and stopping atoms by using lasers. Trapping and cooling the atoms will give other scientists the opportunity to study the fundamental properties of atoms with amazing accuracy and precision.

Who were the **other Nobel prize winners in physics**?

1996 David M. Lee, Douglas D. Osheroff, and Robert C. Richardson
discovery of superfluidity in helium-3

1995 Martin L. Perl
discovery of the tau lepton

Frederick Reines
detection of the neutrino

1994 Bertram N. Brockhouse
development of neutron spectroscopy

Clifford G. Schull
development of the neutron diffraction technique

1993 Russell A. Hulse and Joseph H. Taylor Jr.
discovery of a new type of pulsar

1992 George Charpak
invention and development of particle detectors

1991 Pierre-Gilles de Gennes
discovering that methods developed for studying order phenomena in simple systems can be generalized to more complex forms of matter

1990 Jerome I. Friedman, Henry W. Kendall, and Richard E. Taylor
investigations concerning deep inelastic scattering of electrons on protons and bound neutrons

1989 Norman F. Ramsay
invention of the separated oscillatory fields method and its use in the hydrogen maser and other atomic clocks

Hans G. Dehmelt and Wolfgang Paul
development of the ion trap technique

1988 Leon M. Lederman, Melvin Schwartz, and Jack Steinberger
the neutrino beam method and the demonstration of the doublet structure of the leptons through the discovery of the muon neutrino

1987 J. Georg Bednorz and K. Alexander Müller
discovery of superconductivity in ceramic materials

1986 Ernst Ruska
fundamental work in electron optics, and for the design of the first electron microscope

Gerd Binnig and Heinrich Rohrer
design of the scanning tunneling microscope

1985 Klaus von Klitzing
discovery of the quantized Hall effect

1984 Carlo Rubbia and Simon van der Meer
decisive contributions to the large project, which led to the discovery of the field particles W and Z, communicators of weak interaction

1983 Subramanyan Chandrasekhar
theoretical studies of the physical processes of importance to the structure and evolution of the stars

William A. Fowler
theoretical and experimental studies of the nuclear reactions of importance in the formation of the chemical elements in the universe

1982 Kenneth G. Wilson
theory for critical phenomena in connection with phase transitions

1981 Nicolaas Bloembergen and Arthur L. Schawlow
contribution to the development of laser spectroscopy

Kai M. Siegbahn
contribution to the development of high-resolution electron spectroscopy

1980 James W. Cronin and Val L. Fitch
discovery of violations of fundamental symmetry principles in the decay of neutral K-mesons

1979 Sheldon L. Glashow, Abdus Salam, and Steven Weinberg
contributions to the theory of the unified weak and electromagnetic interaction between elementary particles, including inter alia the prediction of the weak neutral current

1978 Pyotr Leonidovich Kapitsa
basic inventions and discoveries in the area of low-temperature physics

Arno A. Penzias and Robert W. Wilson
discovery of cosmic microwave background radiation

1977 Philip W. Anderson, Sir Nevill F. Mott, and John H. van Vleck
fundamental theoretical investigations of the electronic structure of magnetic and disordered systems

1976 Burton Richter and Samuel C. C. Ting
discovery of a heavy elementary particle of a new kind

1975 Aage Bohr, Ben Mottelson, and James Rainwater
discovery of the connection between collective motion and particle motion in atomic nuclei and the development of the theory of the structure of the atomic nucleus based on this connection

1974 Sir Martin Ryle and Anthony Hewish
research in radio astrophysics; Ryle, observations and inventions, in particular of the aperture synthesis technique; and Hewish, decisive role in the discovery of pulsars

1973 Leo Esaki and Ivar Giaever
experimental discoveries regarding tunneling phenomena in semiconductors and superconductors

Brian D. Josephson
theoretical predictions of the properties of a supercurrent through a tunnel barrier, in particular those phenomena which are generally known as the Josephson effects

1972 John Bardeen, Leon N. Cooper, and John Robert Schrieffer
jointly developed theory of superconductivity, usually called the BCS theory

1971 Dennis Gabor
invention and development of the holographic method

1970 Hannes Alfvén
work and discoveries in magneto-hydrodynamics with fruitful applications in different parts of plasma physics

Louis Néel
work and discoveries concerning antiferromagnetism and ferromagnetism

1969 Murray Gell-Mann
contributions and discoveries concerning the classification of elementary particles and their interactions

1968 Luis W. Alvarez
decisive contributions to elementary particle physics, in particular the discovery of a large number of resonance states, made possible through his development of the technique of using hydrogen bubble chamber and data analysis

1967 Hans Albrecht Bethe
contributions to the theory of nuclear reactions, especially his discoveries concerning the energy production in stars

1966 Alfred Kastler
discovery and development of optical methods for studying Hertzian resonances in atoms

1965 Sin-Itiro Tomonaga, Julian Schwinger, and Richard P. Feynman
fundamental work in quantum electrodynamics

1964 Charles H. Townes, Nicolai Gennadiyevich Basov, and Alexandr Mikhailovich Prokhorov
fundamental work in the field of quantum electronics, which has led to the construction of oscillators and amplifiers based on the maser-laser principle

1963 Eugene P. Wigner
contributions to the theory of the atomic nucleus and the elementary particles, particularly through the discovery and application of fundamental symmetry principles

Maria Goeppert-Mayer and J. Hans D. Jensen
discoveries concerning nuclear shell structure

1962 Lev Davidovich Landau
pioneering theories for condensed matter, especially liquid helium

1961 Robert Hofstadter
pioneering studies of electron scattering in atomic nuclei

Rudolf Ludwig Mössbauer
research concerning the resonance absorption of gamma radiation and his discovery of the effect which bears his name

1960 Donald A. Glaser
invention of the bubble chamber

1959 Emilio Gino Segrè and Owen Chamberlain
discovery of the antiproton

1958 Pavel Alekseyevich Čerenkov, Ilya Mikhailovich Frank, and Igor Yevgenyevich Tamm
discovery and interpretation of the Čerenkov effect

1957 Chen Ning Yang and Tsung-Dao Lee
investigation of the so-called parity laws, which has led to important discoveries regarding the elementary particles

1956 William Shockley, John Bardeen, and Walter Houser Brattain
research on semiconductors and their discovery of the transistor effect

1955 Willis Eugene Lamb, Jr.
discoveries concerning the fine structure of the hydrogen spectrum

Polykarp Kusch
precision determination of the magnetic moment of the electron

1954 Max Born
research in quantum mechanics, especially statistical interpretation of the wavefunction

Walther Bothe
the coincidence method and his discoveries made therewith

1953 Frits (Frederik) Zernike
demonstration of the phase contrast method, especially invention of the phase contrast microscope

1952 Felix Bloch and Edward Mills Purcell
development of new methods for nuclear magnetic precision measurements and discoveries in connection therewith

1951 Sir John Douglas Cockcroft and Ernest Thomas Sinton Walton
pioneer work on the transmutation of atomic nuclei by artificially accelerated atomic particles

1950 Cecil Frank Powell
development of the photographic method of studying nuclear processes and his discoveries regarding mesons made with this method

1949 Hideki Yukawa
prediction of the existence of mesons on the basis of theoretical work on nuclear forces

1948 Lord Patrick Maynard Stuart Blackett
development of the Wilson cloud chamber method, and his discoveries therewith in the fields of nuclear physics and cosmic radiation

1947 Sir Edward Victor Appleton
investigations of the physics of the upper atmosphere and especially for the discovery of the so-called Appleton layer

1946 Percy Williams Bridgman
invention of an apparatus to produce extremely high pressures, and for the discoveries he made therewith in the field of high-pressure physics

1945 Wolfgang Pauli
discovery of the Exclusion Principle, also called the Pauli Principle

1944 Isidor Isaac Rabi
resonance method for recording the magnetic properties of atomic nuclei

1943 Otto Stern
contribution to the development of the molecular ray method and his discovery of the magnetic moment of the proton

1942 No award

1941 No award

1940 No award

1939 Ernest Orlando Lawrence
invention and development of the cyclotron and for results obtained with it, especially with regard to artificial radioactive elements

1938 Enrico Fermi
demonstrations of the existence of new radioactive elements produced by neutron irradiation, and for his related discovery of nuclear reactions brought about by slow neutrons

1937 Clinton Joseph Davisson and Sir George Paget Thomson
experimental discovery of the diffraction of electrons by crystals

1936 Victor Franz Hess
discovery of cosmic radiation

Carl David Anderson
discovery of the positron

1935 Sir James Chadwick
discovery of the neutron

1934 No award

1933 Erwin Schrödinger and Paul Adrien Maurice Dirac
discovery of new productive forms of atomic theory

1932 Werner Heisenberg
creation of quantum mechanics, the application of which has, inter alia, led to the discovery of the allotropic forms of hydrogen

1931 No award

1930 Sir Chandrasekhara Venkata Raman
work on the scattering of light and for the discovery of the effect named after him

1929 Prince Louis Victor de Broglie
discovery of the wave nature of electrons

1928 Sir Owen Willans Richardson
work on the thermionic phenomenon and especially for the discovery of the law named after him

1927 Arthur Holly Compton
discovery of the effect named after him

Charles Thomson Rees Wilson
method of making the paths of electrically charged particles visible by condensation of vapor

1926 Jean-Baptiste Perrin
work on the discontinuous structure of matter, and especially discovery of sedimentation equilibrium

1925 James Franck and Gustav Hertz
discovery of the laws governing the impact of an electron upon an atom

1924 Karl Manne Georg Siegbahn
discoveries and research in the field of X-ray spectroscopy

1923 Robert Andrews Millikan
work on the elementary charge of electricity and on the photoelectric effect

1922 Niels Bohr
investigation of the structure of atoms and of the radiation emanating from them

1921 Albert Einstein
discovery of the law of the photoelectric effect

1920 Charles Édouard Guillaume
discovery of anomalies in nickel steel alloys

1919 Johannes Stark
discovery of the Doppler effect in canal rays and the splitting of spectral lines in electric fields

1918 Max Karl Ernst Ludwig Planck
discovery of energy quanta

1917 Charles Glover Barkla
discovery of the characteristic Röntgen radiation of the elements

1916 No award

1915 Sir William Henry Bragg and Sir William Lawrence Bragg
services in the analysis of crystal structure by means of X-rays

1914 Max von Laue
discovery of the diffraction of X-rays by crystals

1913 Heike Kamerlingh Onnes
investigations on the properties of matter at low temperatures which led, inter alia, to the production of liquid helium

1912 Nils Gustaf Dalén
invention of automatic regulators for use in conjunction with gas accumulators for illuminating lighthouses and buoys

1911 Wilhelm Wien
discoveries regarding the laws governing the radiation of heat

1910 Johannes Diderik van der Waals
work on the equation of state for gases and liquids

1909 Guglielmo Marconi and Carl Ferdinand Braun
contributions to the development of wireless telegraphy

1908 Gabriel Lippmann
method of reproducing colors photographically based on the phenomenon of interference

1907 Albert Abraham Michelson
optical precision instruments and the spectroscopic and metrological investigations carried out with their aid

1906 Sir Joseph John Thomson
theoretical and experimental investigations on the conduction of electricity by gases

1905 Philipp Eduard Anton Lenard
work on cathode rays

1904 Lord John William Strutt Rayleigh
investigations of the densities of the most important gases and for his discovery of argon in connection with these studies

1903 Antoine-Henri Becquerel
discovery of spontaneous radioactivity

Pierre Curie and Marie Curie, née Sklodowska
joint research on the radiation phenomena discovered by Professor Henri Becquerel

1902 Hendrik Antoon Lorentz and Pieter Zeeman
research into the influence of magnetism upon radiation phenomena

1901 Wilhelm Conrad Röntgen
discovery of the Röntgen rays

Who was the **first American** to win the Nobel prize in physics?

In 1907, for the development of extremely precise measurements for the velocity of light and his work on optical instruments, German-born Albert A. Michelson—a naturalized U.S. citizen—won the Nobel prize in physics. It would be fourteen years until Albert Einstein, a German who would one day become an American citizen, won the Nobel prize in physics for his work on the photoelectric effect.

What country has produced the **most winners** of the Nobel prize in physics?

Since 1901, when the Nobel prize was first awarded, the United States has had more Nobelists in physics than any other country, although initially it took six years before a U.S. citizen won a Nobel prize in physics.

Who were the **two women** to win the Nobel prizes in physics?

In 1903, Marie Curie was the first woman to win the Nobel prize in physics. She was awarded the prize with her husband, Pierre, and with Antoine Becquerel for their discovery of over forty radioactive elements and other breakthroughs in the field of radioactivity.

In 1963, Maria Goeppert-Mayer became the second woman and the first and only American woman to win the Nobel prize in physics.

What is meant by the phrase **"motion is relative to one's perspective"**?

Everything in the universe is in motion. The Earth is rotating on its axis, and orbiting the sun along with other planets in our solar system; our solar system itself is moving among other stars within our galaxy, which in turn moves among other galaxies in the universe. "Standing still," then, is technically impossible. When motion is discussed, it must be described as "relative or compared to something else." Unless the frame of reference is stated, motion is usually assumed to be relative to a person standing on the surface of the Earth.

Even relative to the Earth, however, an object may seem to be at rest to one person, while to someone else it may appear to be moving. Take, for example, a person reading *The Handy Physics Answer Book* in a car driving down the road. From the reader's point of view, the book is not moving and is at rest relative to the reader. However, for an observer standing on the side of the road, that book, as well as the person reading the book inside the car, is moving. Depending upon the point of view, the book has two different motions and therefore must be described from a particular perspective.

SPEED, VELOCITY, AND ACCELERATION

Aren't **speed and velocity** considered to be the **same thing**?

Speed and velocity are often used interchangeably; to physicists, however, speed and velocity differ by the specification of direction. Speed is the distance an object travels during a particular time interval. If a vehicle travels 100 km (kilometers) in one hour, the vehicle's speed is 100 km/h. Velocity's definition is the same as speed, except that in addition to how fast the object travels, velocity also defines the direction in which the object moves. For example, the velocity of the same vehicle might be 100 km/h east. Therefore, a turning vehicle may be moving at a constant speed, but not a constant velocity, because its direction is changing.

When speeds are listed as **meters per second** (m/s), what does that mean in **miles per hour**?

In this book most of the speeds are listed in meters per second (m/s). The following list is a chart that compares some of the speeds in meters per second to miles per hour (mph):

Meters per Second	Miles per Hour
5 m/s	11.2 mph
10 m/s	22.3 mph
15 m/s	33.5 mph
20 m/s	44.6 mph
25 m/s	55.8 mph
30 m/s	66.9 mph
35 m/s	78 mph

What is **acceleration**?

Acceleration is defined as how quickly an object changes its velocity; that is, the change of velocity divided by the change of time (the difference in time over which the velocity changed). There are three methods of acceleration: speeding up (accelerating), slowing down (decelerating), and turning (changing direction).

What is the ultimate speed limit?

The fastest speed possible is the speed of light. This "ultimate speed limit," as determined by Einstein in his theories on Relativity, is 3×10^8 m/s (meters per second). According to Einstein, if anyone could actually ever achieve the speed of light. time would stop for the person traveling at that speed. (For more information on Einstein's theories of Relativity, refer to the DEEP THEORIES chapter.)

A vehicle that changes its velocity from 0 m/s (meters per second) to 10 m/s in one second has an acceleration of 10 m/s^2, whereas an object traveling at 10 m/s that stops in one second has an acceleration of -10 m/s^2. The negative sign represents a deceleration, or acceleration in a negative direction.

NEWTON'S THREE LAWS OF MOTION

INERTIA

What is Newton's **First Law of Motion**?

The Law of Inertia is the first of Newton's three laws that govern motion. Inertia is an object's resistance to changing its motion. The Law of Inertia says that if an object is at rest, it wants to stay at rest; if an object is in motion, it keeps a constant velocity, until something exerts a force on it to change its motion. How much inertia an object has is determined by the object's mass; mass is an artificial measurement of inertia.

For example, a puck on a low-friction air hockey table will keep moving in a straight line with a constant speed until a force acts upon it. That force could be the wall of the hockey table, a person, or the friction from the air resistance. However, until something pushes or pulls on the puck, it will move with a constant velocity.

Why is it important that **linebackers** in football have a lot of **mass**?

Inertia is dependent upon mass, so the more mass a linebacker has, the harder it is to push the linebacker out of the way. The opposite is needed for a wide receiver; they need to be able to accelerate quickly, so if a receiver has less mass, it is easier to run quick football patterns that require many changes in motion.

Why are **headrests** necessary in all vehicles?

A headrest is needed for the same reason as seat belts: to increase safety. Headrests are not designed for "resting" or comfort, but to prevent the head from snapping backwards when a vehicle is struck from behind. When rear-ended, a vehicle and its seat exert a force on a person's body, accelerating them forward. The seat does not, however, push the person's head forward, so—according to the Law of Inertia—the head will keep its original motion relative to the ground. This means that the head will snap backward relative to the vehicle. A vehicle equipped with a headrest helps push the head forward with the rest of the body, preventing serious neck injuries that might result from the "snapping."

Force

What is a **force**?

Isaac Newton defined force as a push or pull that when applied to a mass causes the object to accelerate. The formula "Force = mass × acceleration" or "F = ma" is known as Newton's Second Law of Motion.

If the mass of an object is held constant, the larger the force that is applied, the greater the acceleration. Take, for example, a compact car

How do seat belts save lives?

A seat belt is one of the most important personal safety devices in an automobile. The seat belt provides a force that changes the motion of a person when he or she is in an accident or comes to an abrupt stop. If the vehicle stops suddenly, the person's body still wants to move forward at the same rate at which the vehicle was moving prior to the collision. The seat belt locks to apply a force backward on the person, and instead of hitting the dashboard or crashing through the windshield, the person stays in his or her seat and hopefully avoids injury.

that has a small engine that can exert only a small force. That small force allows for only a small rate of acceleration. However, the same compact car with a more powerful engine could apply a greater force and accelerate at a higher rate than before. When the mass of an object is kept constant, the force and acceleration are directly proportional to each other.

If, instead, the force is held constant, the larger the mass or inertia of the object, the smaller the rate of acceleration. Take again the compact car that has a small amount of mass and an engine that applies a specific force, which causes the car to accelerate. If that same engine were placed inside a Mack truck, the inertia of the truck would allow a much smaller rate of acceleration than the compact car. When the force is kept constant, the mass and acceleration are indirectly proportional to each other.

Interacting Forces

Why does every action force have an **equal and opposite reaction** force?

Newton's Third Law of Motion states that equal and opposite forces occur when two or more objects interact. For example, when a softball

Why do tires have treads?

In order for tires to turn and move a car forward, they need to have friction between the rubber tire and the road. If there were no friction, the tires would simply spin much like tires spin on low-friction ice. More rubber on the road results in more friction, which allows for better control of the vehicle. Although treads reduce the contact area between the tire and the road—thus reducing friction—treads contribute to safer road handling.

The reason conventional tires have treads is to displace and redirect the water from between the road surface and the underside of the tire. Since water serves as a lubricant on wet road surfaces, the tire would not be able to establish enough friction with the road if such a lubricant was present. It is important to occasionally check your tires' tread levels to make sure they can still be effective on slippery surfaces.

hits a bat, the bat also hits the softball. Newton realized that for every action force—the softball hitting the bat—there is an equal and opposite reaction force—the bat hitting the softball. The magnitude of the force is the same; however, according to Newton's Second Law, the object with less mass (and therefore less inertia) will accelerate more. The softball, since it is less massive than the bat and the person holding the bat, will accelerate more than the bat.

How does Newton's **Third Law** explain how we **walk**?

Newton's Third Law of Motion states that forces occur in equal and opposite pairs. When a person walks, her foot pushes against the Earth, but the Earth, as a reaction force, pushes her back. The person and the Earth both apply equal amounts of force, but because the Earth is so massive and the person has so little inertia in comparison, the walker moves or accelerates much more than the Earth.

Treadless racing tires.

FRICTION

What is **friction**?

Whenever two or more objects interact, their irregular surfaces slide and scrape against one another, impeding the motion of the objects. Friction is this force that opposes the motion. The amount of friction between two objects depends upon how smooth or rough the surfaces are, and the pressure between them. Since no surface is perfectly smooth, all objects experience friction when interacting with other objects.

Why don't **racing cars** have treads on their tires?

Racing cars only race on dry days, so they have little need for redirecting water away from the underside of the tire. By not having treaded tires, they can apply more rubber to the roadway; the increased friction creates better grip and provides safer handling.

What is the unit that **measures the frictional characteristics** of an object?

An object's coefficient of friction, "μ," a Greek letter pronounced "mu," is how physicists and engineers measure the amount of smoothness or roughness between materials. The higher the value for μ, the greater the roughness.

What is the difference between **static friction and kinetic friction**?

Static friction is the friction between two objects at rest; kinetic friction is that between two objects in motion. Objects have greater frictional forces while at rest than when they are moving. Take, for example, the task of moving a large, massive crate that is at rest on a concrete floor. A person must apply a large force in order to overcome the static frictional forces between the floor and the crate, which are at rest. Just as it starts to slide, the crate, according to the Law of Inertia, wants to keep moving. Since the crate is already moving, the force needed overcome the kinetic friction is a bit less than the force to overcome static friction.

How can friction be **reduced**?

Friction can be reduced by making the surfaces smoother or by reducing the pressure between the two materials. Sometimes a more practical way to reduce friction is to introduce a lubricant. A lubricant is a fluid that fills in the scrapes, bumps, and valleys to reduce the amount of scraping between the surfaces. As a result, lubricant allows objects to slip and pass over each other with much less friction.

How can **ball bearings** reduce the amount of friction?

Ball bearings are small steel balls located in a circular track between an axle and a wheel. As the wheel turns, the ball bearings reduce the amount of friction by rolling in the circular track around the axle. The rolling motion creates less of a frictional force than the spinning or sliding friction that would otherwise be present when the wheel would rub against the axle because in a rolling motion, there is minimal surface

Is friction always bad?

There are good and bad types of friction, depending upon what needs to be done. For example, friction between car engine parts causes kinetic or moving energy to change to heat, which results in a less efficient vehicle. However, if it weren't for friction, it would be impossible for tires to push against the pavement.

contact between the objects. Secondly, ball bearings and track are made with very smooth, hard surfaces—usually smoother than the surfaces of the wheel and axle. Finally, the bearings are lubricated, often with a heavy grease. The combination of round bearings, smooth surfaces, and lubricants significantly reduces friction.

How does friction **decrease efficiency** in machines?

The scratching and scraping that occurs when two objects rub against each other changes the moving—or kinetic—energy into heat—or thermal—energy. Hence, because energy is taken away from the moving parts and transferred to the surrounding elements as heat, it reduces the amount of work that the machine can perform, and increases wear and the chance of mechanical breakdown. A perfect machine would have zero energy changed to heat as a result of frictional forces. Although there are ways to reduce the amount of friction in machines, there is no such thing as a frictionless or ideal machine.

Is there a place where friction **does not exist**?

Friction exists throughout the universe. Although there are many ways that friction can be reduced, it cannot be eliminated. The closest thing to a frictionless environment would be in space. Since space is a vacuum, there would be no air resistance to slow down the motion of objects, and as long as there were no major gravitational forces in the vicinity, there would be no natural pressure between the surfaces of two objects. However, if the objects touched, the contact between the surfaces would create friction. Therefore, as long as there is no contact between objects such as stars, planets, dust, or gas, space can be considered a nearly frictionless environment.

FREE FALL

What does it mean to be in **free fall**?

Free fall occurs when an object is pulled down by the gravitational force of the Earth or any other large gravitational body. However, objects descending to the Earth are not usually in true free fall, because frictional forces—such as air resistance—hinder an object's acceleration toward the ground. The only way that true free fall can be achieved is in a vacuum, where there is no air resistance.

How **fast** does one fall in free fall?

If air resistance is not taken into consideration, an object accelerates downward at a rate of 9.8 m/s^2 (meters per second squared) or 32 ft/s^2 (feet per second squared). The freely falling object increases its speed as it falls by 9.8 m/s for every second it falls. The following chart illustrates the speed (where velocity is equal to gravity × time) and the distances for falling objects:

Time (sec)	Speed (m/s) $v = gt$	Distance (m) $d = .5\ gt^2$	Speed (ft/s) $v = gt$	Distance (ft) $d = .5\ gt^2$
1	9.8	4.9	32	16
2	19.6	19.6	64	64
3	29.4	44.1	96	144
4	39.2	78.4	128	256
5	49.0	122.5	160	400
6	58.8	176.4	192	576
7	68.6	240.1	224	784

Skydivers in free fall.

Time (sec)	Speed (m/s) $v = gt$	Distance (m) $d = .5\ gt^2$	Speed (ft/s) $v = gt$	Distance (ft) $d = .5\ gt^2$
8	78.4	313.6	256	1024
9	88.2	396.9	288	1296
10	98.0	490.0	320	1600

What is **terminal velocity**?

Terminal velocity is the maximum speed that a person or object can achieve while falling toward the Earth. The term is used in skydiving as the "top speed" a person achieves during a fall. Without air resistance, a person would theoretically fall faster and faster, accelerating at a rate of 9.8 m/s^2 (meters per second squared) until he strikes the Earth. But in reality, there is a rather large air resistant force when jumping from an airplane. Terminal velocity occurs when the air resistant force upward on the diver equals the downward force of gravity acting on the diver. When the two forces are equal, the person stops accelerating and falls to the Earth at a constant velocity.

How **fast** can a person **fall**?

The maximum or terminal velocity for the average skydiver is about 150–200 km/h (kilometers per hour), or 93–125 mph (miles per hour). Terminal velocity of a skydiver is dependent upon the person's weight, as well as how large or aerodynamic the person is.

How does a **parachute** work?

A parachute is deployed by a skydiver to increase the amount of air resistance and slow down the fall. Once the parachute is deployed, the air resistant force is larger than the force of gravity, so the diver slows down dramatically. Within a second or so, terminal velocity with the open parachute is achieved, allowing the person to land on the ground with a speed of about 15–25 km/h (kilometers per hour), or 9–16 mph.

PRESSURE

What is **pressure**?

Pressure is defined as the amount of force that is applied to a specific area. The formula for pressure is force divided by area, so the more concentrated and stronger the force, the greater the pressure; pressure is lower with a greater area and/or a weaker force. Pressure is typically measured in pounds per square inch (psi) in the English system of measurement, and in Pascals (Pa) or newtons per square meter in the metric system.

Why do needles, pins, nails, spikes, and arrows have **pointed tips**?

The purpose behind these objects is to easily penetrate a particular surface; that is, to puncture with a minimal amount of pressure. Since pressure is defined as force divided by area, and the pointed end has a very small surface area, only a small amount of force needs to be applied by the user to create a large amount of pressure.

Why would it be dangerous to drop a penny off the Empire State Building?

A penny falling from the top of the Empire State Building would prove extremely dangerous to a person walking at ground level. The penny would accelerate until it reached a terminal velocity of about 175 km/h (kilometers per hour), or 109 mph (miles per hour). If the falling penny were to hit someone, a very large force would result, possibly breaking the person's skin or even resulting in a serious head injury.

Why is it difficult to walk in grassy areas with **high heels**?

This answer may be obvious to those who wear high heels, but for those who don't, high heels tend to sink in when walking on soft ground. This is because the weight of the person is concentrated over a small area—the heel—which creates high pressure, pushing the heels into the earth. This phenomenon is not limited to heavier high-heel wearers, either; a 120-pound person could exert almost 2000 lbs/in^2 (pounds per square inch) if she balanced on one foot. The same principle applies to cleats, but to the wearer's advantage. Athletes in soccer, baseball, and football use the high pressure from their cleats to penetrate the ground, achieving greater traction and preventing their feet from slipping.

Can a person really lie down on a **bed of nails**?

A bed of nails is a board that has many nails sticking up from the base. When a person lies on the bed, his or her weight is distributed over so many nails that the pressure from each nail is not enough to pierce the skin. The keys are using lots of nails and lots of body surface area. It is much easier to lay down on hundreds of nails than just one—the greater the number of nails, the lesser the amount of pressure on each nail. Decreasing the surface area—for example, by sitting up instead of laying down—would also increase the pressure on the nails. Although the prin-

A bed of nails.

ciple behind the bed of nails is fairly simple, this trick should, indeed, not be tried at home.

How do **ice skates** work?

When standing on ice skates, a large amount of pressure is present between the metal blade and the ice, because the weight of the skater is concentrated over a very small area. If the person were wearing shoes, the weight would be spread out over a larger area, resulting in a lower pressure. The high pressure causes the melting point of the ice to decrease, in turn causing a small amount of ice directly under the skate to melt. When the skate is put into motion, it does not scratch or rub against the solid ice, but instead glides or slips across the water that the increased pressure created. When the skate leaves the melted area, the water freezes again due to the freezing temperature of the surrounding ice. This same phenomenon can be seen by placing a string with small weights tied to each end over an ice cube in the freezer. If allowed to stand overnight, the string will melt a little bit of the ice, sink in, and be imbedded in the cube by the next morning.

MASS VS. WEIGHT

What is the difference between **mass and weight**?

Although weight is dependent upon mass, they are quite different. Mass is the measurement of an object's inertia, which is dependent upon the amount of matter that an object possesses. Mass never changes for an object as long as it has not lost or gained matter. Mass is measured in kilograms (kg) in the metric and SI systems of measurement, and in slugs in the English system.

Weight, however, is dependent upon how much mass an object has, but can change when the object moves from one level of gravitational force to another, such as from the Earth to the moon. Weight is the product of the mass of an object and the acceleration caused by gravity; near the surface of the Earth, that would be $W = mg$ (weight equals mass times gravity, where here gravity means the gravitational acceleration of the Earth's gravitational field). Weight is expressed in newtons (N) in the metric and SI systems, and in pounds (lbs) in the English system.

A person with a mass of 50 kg (3.5 slugs) multiplied by the acceleration due to gravity (9.8 m/s^2) would weigh 490 N (newtons), or 110 lbs, on Earth. However, if that same person were on the moon, where the gravitational acceleration is 1/6th that of Earth, then the person would weigh only 80 N (18 lbs). Finally, if the person were in empty space, free from gravitational fields from other planets, then the person would weigh 0 N, yet still have a mass of 50 kg.

Since the moon has less gravity than the Earth, why is it just as **difficult to move objects on the moon**?

To *lift* an object on the moon would be easier than lifting the same object on Earth, because the object would weigh less on the moon, due to the moon's lower rate of gravitational acceleration. However, if an object needed to be moved *from side to side,* it would be just as difficult on the moon as it would be on Earth. This is because gravity governs only vertical motion—lifting and lowering—and has no control over horizontal motion—pushing or pulling.

How much would a 110-pound person weigh on **Mars**?

A person weighing 110 lbs (50 kg of mass) on Earth would weigh...

Planet	Acceleration Due to Gravity	Weight (newtons)	Weight (pounds)
Mercury	3.72 m/s²	186.0 N	42 lbs
Venus	8.92 m/s²	446.0 N	100 lbs
Earth	9.80 m/s²	490.0 N	110 lbs
Mars	3.72 m/s²	186.0 N	42 lbs
Jupiter	24.80 m/s²	1240.0 N	278 lbs
Saturn	10.49 m/s²	525.0 N	118 lbs
Uranus	9.02 m/s²	451.0 N	101 lbs
Neptune	11.56 m/s²	578.0 N	130 lbs
Pluto	0.29 m/s²	14.5 N	3 lbs

When Apollo Astronaut David Scott dropped a **feather and a hammer on the moon,** which hit the ground first?

While on the moon, David Scott conducted one of the most famous demonstrations of free fall. Scott proved through experimentation what Galileo stated 300 years before. In his simple experiment, Scott held a hammer in one hand, and a feather in the other. He dropped the two, and they both fell and hit the ground at the same time. This proved what others had previously confirmed on Earth: that Galileo was correct in stating that all objects, in the absence of air resistance, fall at the same rate. Since there is no air resistance on the moon, the demonstration worked perfectly.

How far away would one have to travel from the Earth to experience **no gravitational effects**?

According to Newton's Law of Universal Gravitation, an object of mass will experience a gravitational attraction to the Earth, no matter where the object is in the universe. However, after achieving a particular distance from the Earth, that force would be too small to notice. At approximately 2,640 km (kilometers) or 1,640 miles, the gravitational pull would be 50% of what it would be on the surface of the Earth. At a dis-

If someone could be placed in the center of the Earth, what would she weigh?

If we could situate ourselves in the center of the Earth, we would be weightless. People are gravitationally attracted to the Earth because of its mass. If a woman were placed directly in the center of the Earth, then there would be equal amounts of mass above, below, in front, behind, to the left, and to the right of her. She would be attracted to all this mass, but the mass would be pulling her equally in all directions. The result would cause the gravitational forces to cancel each other out, resulting in a weightless state!

tance of 574,000 km or 357,000 miles, the force would be 1% of the gravitational pull, or virtually weightless. That distance is equal to nine Earth radii away from the surface of the Earth, or four and a half times the Earth's diameter.

GRAVITY AND GRAVITATIONAL INTERACTIONS

Who discovered **gravity**?

Physicists such as Aristotle and Galileo developed theories and experiments to understand why objects fall toward the Earth. It was Isaac Newton, however, who finally understood gravity. Newton realized that all objects in the universe that have mass are attracted toward one another. Newton developed calculus in order to prove this theory and eventually define the Law of Universal Gravitation.

Why do objects fall?

Everything on Earth is attracted to the Earth because there is a gravitational attraction between all objects of mass. Therefore, the Earth, which is extremely massive, and an apple, that also has mass, would be attracted by a gravitational force that would cause both objects to accelerate toward each other. The reason why the apple is moving more than the Earth is because the Earth has much more mass than the apple, thus more inertia; as a result, the Earth is less willing to move toward the apple.

Did a falling apple cause **Isaac Newton** to discover gravity?

According to Isaac Newton, his fascination with gravity began one fall day in Woolsthorpe, England, in 1665. He had left Cambridge to escape the bubonic plague, which had killed thousands of people throughout Europe. To hide from the distractions of the town and concentrate on his studies, Isaac would often spend time in a nearby apple orchard. It was there that an apple fell at his feet and got the young Newton thinking about gravity. He wondered how objects, like the apple, could be attracted to the Earth even though they were not in contact with the Earth. Newton would later prove that the same gravitational force that pulls an apple to the ground would be the same force that causes the moon to orbit the Earth.

What is the **Law of Universal Gravitation**?

Gravity, as stated by Newton, is a force that attracts every particle of mass to every other particle of mass in the entire universe. That means that a person here on Earth is attracted to every other person, object, planet, and star throughout our universe. However, the reason why we do not gravitate toward these objects is that the gravitational force between Earth and ourselves is much stronger than any gravitational force between other people, buildings, distant planets, or stars.

Sir Isaac Newton.

What **factors** determine the **gravitational attraction** between two objects?

Two factors determine how strong a gravitational pull is between two objects. The greater the masses of the objects, the greater the gravitational force between them. The second variable is distance. Newton concluded that the gravitational attraction decreases as the square of the distance between the two masses increases. The greater the distance between the two masses, the less the gravitational pull.

How did the Law of Universal Gravitation help in the **discovery of Neptune**?

Newton's Law of Universal Gravitation explains how much gravitational attraction exists between the sun and Uranus. In 1846, after noticing peculiar perturbations (fluctuations) in the orbit of Uranus, a French astronomer named Urbain Leverrier mathematically calculated the position of an eighth planet that was responsible for the perturbations in Uranus' orbit. Within the same year, Johann Galle, a German astronomer, observed that eighth planet—Neptune—and defined its position in space less than a degree off of its theoretical position.

What other planet was discovered as a result of **Uranus' perturbations**?

Astronomers in 1930 at the Lowell Observatory in Arizona made a significant observation that led to the discovery of Pluto, the ninth planet in our solar system. American astronomer Percival Lowell observed that Uranus, during certain sections of its orbit, made strange fluctuations and movements that were not consistent with a normal, elliptical orbit. Although Neptune was the popular explanation of this phenomenon, Lowell hypothesized in 1905 that the perturbations must have been caused by an additional object or planet beyond Neptune's orbit. Although Lowell was not able to find the planet causing the disturbance, twenty-five years later, astronomer Clyde Tombaugh observed the ninth planet extremely close to where Lowell had originally predicted the location of Pluto.

Why is **Pluto** not always the outermost planet in our solar system?

Although Pluto is considered to be the outermost planet in our solar system, as of the time this book was published, it is actually a bit closer to the sun than Neptune. The reason is that Pluto's orbit is extremely elliptical and during its perigee, when Pluto is at its closest position to the sun, it is actually closer to the sun than Neptune. Although Pluto is the outer planet for most of its orbital period, Neptune will be the farthest planet from the Earth from January 23rd, 1979, to March 15, 1999.

If **Pluto** is not always the outermost planet, and its path crosses that of **Neptune,** then will the two planets **collide**?

Although Pluto revolves around the sun in an orbit 5.9 billion kilometers in length for approximately 250 years in duration, the likelihood of collision would be quite small for such large orbits. However, even the possibility of such an event would be impossible because the path of Pluto is inclined at an angle of 17.2° and never truly crosses Neptune's path. In many two-dimensional orbital diagrams, it would appear that they cross, but in three-dimensional diagrams, it is clear that they actually do not.

Could there be another planet **beyond Pluto**?

The discovery of Pluto did not completely explain Uranus' perturbations; it was determined that Pluto's mass and distance from Uranus were not significant enough to cause such perturbations. As a result, the search for a possible tenth planet, "Planet X," continues.

TIDAL FORCES

What causes **tides**?

Tidal forces are caused by the gravitational interaction between the moon and the Earth. The gravitational force from the moon causes the

Canada's Bay of Fundy at low tide (see page 50).

water in the oceans to be slightly attracted to the moon. The section of the Earth that is closest to the moon experiences the larger gravitational force, and as a result, accumulates a vast amount of water in this location. This represents a high tide. The second location of a high tide is located on the opposite side of the Earth, the section farthest away from the moon, caused by the difference in gravitational pull between the near and far side of the Earth from the moon.

The low tides occur in areas that are not closest or farthest away from the moon. There are two low tides, also occurring at opposite ends of the Earth. The reason for the low tide is that a great deal of water has been displaced and moved to the high-tidal areas.

Since the sun exerts a larger gravitational pull on the Earth than the moon does, why is the **sun not responsible for tides**?

The sun does play a part in tidal forces, but a minor one. Although the gravitational pull from the sun is greater than the gravitational pull from the moon, the difference in gravitational pull on the different regions of the Earth is what causes the tidal surges. The difference in pull from the sun is not very different on the near and far sides of the Earth; therefore, the sun's effect on tides is minimal.

Can tides be **predicted**?

Tides are very predictable. In fact, tide tables can be purchased and are extremely important when navigating in shallow waters. As long as the location, date, and time are known, the depth of water can be determined, and knowing the difference between low and high tide is crucial for a successful ocean voyage.

Tides never occur at the same time day after day. Since tides rely mostly on the position of the moon, and the moon orbits the Earth every 24 hours, 50 minutes, and 28 seconds, tides occur a bit later day after day.

What causes **high tides** to be unusually high and **low tides** to be unusually low?

Although the sun only plays a minor role in tidal forces compared to the moon, when the Earth, moon, and sun align, the gravitational differences between the near and far sides of the Earth are greater than normal. As a result, the tidal forces on the oceans are abnormally large and result in large high tides and very low low tides. These tides are called spring tides, and can be observed when there is a full or new moon.

When the moon and sun are not aligned, a half moon is seen in the night sky, and the gravitational difference on the opposite sides of the Earth are not as great, resulting in smaller high tides called neap tides.

What is the difference between a **flood current** and an **ebb current**?

Flood and ebb currents occur as a result of the tidal forces. After an area has experienced a high tide, the water needs to flow toward a neighboring area that is experiencing a high tide. The outward flow or the lowering of water is called the ebb current. When the water rises for the next

Why don't lakes usually experience significant tides?

Self-contained lakes (lakes without large rivers emptying out to the ocean) do not experience noticeable tidal forces. In order for tides to occur, there must be a significant difference in gravitational pull on two different sections of the lake. Since there is no lake that extends from one side of the globe to the other, the difference in gravitational force is usually not large enough to experience a noticeable tidal bulge.

high tide it is called the flood current. The time for each current is six hours and twelve minutes, the time between each low and high tide.

Where are the **largest tide fluctuations**?

One of the highest tidal fluctuations in the world occurs in Canada's Bay of Fundy. The large tidal fluctuations are not due to a stronger tidal force in this area, but instead to a funneling effect when water from two sections of the bay merge together. The difference is so great that the water depth between low and high tides can vary as much as 18 meters or 60 feet. The dramatic difference means that a boat sailing 60 feet above the bay floor may hit bottom just six hours later. Mariners in the Bay of Fundy need to pay close attention to the tide tables for the bay.

Have engineers ever used tides to generate **hydroelectric power**?

Waterfalls and dams have long been used to turn turbines to generate electrical energy. Although tides cause currents that could turn turbines, tidal currents generally are gradual and do not produce large enough currents to make a hydroelectric plant economical. However, on the Rance River, located in northern France, the tidal surges can be as great as 8.5 meters or 28 feet between the ebb and flood tides. The large amount of water flowing in and out of the mouth of the Rance River produced a large enough current to warrant the construction of a hydroelectric plant in 1966.

IMPULSE AND MOMENTUM

What is **momentum**?

Momentum is a value that describes the amount of inertia and motion an object has, and is derived by the formula (where "P" is momentum) $P = mv$, or mass times velocity. A compact car traveling at 20 miles per hour has much less momentum than a large truck traveling at the same velocity because the truck has more mass.

Momentum is often discussed in sports. Football players quite frequently refer to how much momentum they have. It is advantageous in football not only to have a lot of mass, but a lot of velocity as well. The more momentum a player has, the more difficult it is to stop the player.

What is an **impulse**?

An impulse describes how a change in momentum occurs. In order to change an object's motion or momentum, a force needs to be applied to the object for a period of time. The amount of force and the length of time it is applied will control how much effect the impulse has on the objects.

Why do people **bend their legs** when landing after a jump?

When someone jumps, they typically bend their legs on the landing. An impulse is experienced, and the amount of force and length of time the force was applied determine how much the landing will hurt the legs. Therefore, by bending his or her legs, the jumper will increase the amount of time in the collision by gradually slowing down instead of immediately bringing the body to a halt. This results in a decrease of force between the legs and the ground. If the legs were kept stiff and straight, the time to stop the motion would have been very short, resulting in a very large force and painful experience. Regardless of how the landing takes place, the impulse will be the same; however, by changing the variables that define impulse, the jumper will be able to walk away from the landing, instead of experiencing serious hip and leg injuries.

Why does a **catcher's mitt** tend to have **more padding** than a traditional baseball or softball glove?

The reason for padding baseball gloves is to decrease the force felt by the ball-glove collision by increasing the amount of time it takes for the ball to come to rest. By increasing the time in the collision, the force is reduced to a tolerable level. A catcher's mitt has more padding than the conventional glove because a baseball thrown from a pitcher typically travels faster and has much more momentum than a ball hit to an outfielder. Again, the padding increases the time in the collision, which reduces the force.

What are some other ways that forces are **gradually decreased** to avoid dangerous collisions?

Whenever we experience some sort of collision, we unknowingly play with the two variables, force and time, that define the impulse we experience when in the collision. By increasing the time a force is applied to us when in a collision, we in turn reduce the force exerted on our bodies. The following examples are meant to illustrate a gradual decrease in force to prevent injury, by increasing time.

Trained boxers, when hit in the face, are trained to "ride with the punches." This means that if someone strikes a boxer in the face, the boxer should allow his head to move backward with the punch. This increases the time that the hand is in contact with one's face, but reduces the magnitude of the force, making the punch less effective than if the head and neck were kept stiff.

Another example of increasing the time in a collision to reduce force is found in automotive technology. "Crumple zones" are points in the

Cleveland Indian's Al Lopez with padded catcher's mitt.

How does the air bag expand?

Air bags expand from the steering wheel or dashboard when a sensor, which has been triggered due to the sudden impulse or change in momentum caused by a collision, sends a brief pulse of electricity to a heating element. The heating element causes a chemical reaction with a propellant that fills the air bag with nitrogen gas in approximately 1/20th of a second (0.05 sec). This gives the air bag enough time to expand before the person hits the bag. After 0.3 seconds have elapsed, the collision should be complete and the airbags empty.

frame of automobiles that are designed to "give," or crumple, when in an accident. Since the frame crumples, it allows the vehicle and the person inside the vehicle to come to a more gradual stop than if the frame did not "give."

AIR BAGS

What function does an **air bag** serve in an **automobile collision**?

Whenever a vehicle experiences a front-end collision, the automobile experiences an impulse or change in momentum. The driver and passengers in the car have inertia and will continue to travel forward until the dashboard, seatbelt, or air bag forces them to stop. The force exerted on the passengers when hitting the dashboard could be quite serious and would be devastating if travelling at typical highway speeds.

The function of an air bag is to provide a cushion-like effect to gradually bring the passengers to a halt, instead of allowing them to hit the dashboard or windshield with a tremendous force. Through the use of an airbag, the time to stop the passengers is extended, resulting in a small-

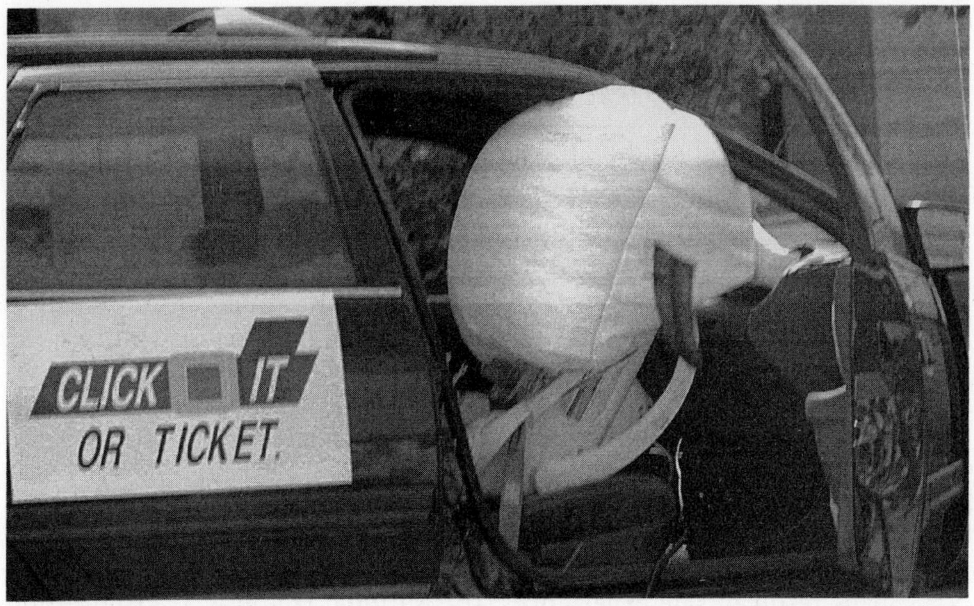

An air bag completely covering a baby dummy in a child seat placed in the front seat of a car, illustrating how *not* to restrain a child.

er force on the individual's body. This safety feature, along with the seatbelt, has helped prevent death and serious injury for many motorists.

When were air bags **first introduced**?

Although air bags have gained a great deal of popularity within the past several years, the idea of an air cushion was proposed back in the 1960s. The first air bags were designed to accommodate unbelted male drivers, 170 lbs and 5 feet 9 inches tall. In the early '70s, the automobile industry debated whether or not it should install air bags in vehicles. Many companies believed that children and short adults might be at risk for injury from the air bags and instead felt that more emphasis should be placed on having motorists wear seat belts.

Why are air bags still surrounded by **controversy**?

Although airbags are standard equipment in many new vehicles, the controversy around the effectiveness of air bags is still brewing. Today's air bags are designed to strike the average seat-belted male in the mid-section of the body. An air bag (which expands at a rate of 150 miles per hour) can be quite dangerous if the bag strikes shorter individuals in the face or neck.

Several rules should be followed to help prevent air bag injuries for young children. First, never place children, whether in a car seat or not, in the front seat of an automobile. Secondly, keep all children in the back seat away from the air bags and make certain they are buckled. If the child absolutely needs to sit in the front seat, make sure there are sufficient restraints on the child and slide the seat as far away from the air bag as possible.

How could air bags be **safer** for the general public?

Although the vast majority of collisions involving cars equipped with air bags result in safe experiences, the National Highway Transportation Safety Agency has made several suggestions on how to improve the already impressive survival rate. The first proposal would be for car manufacturers to provide the option of temporarily deactivating the air bag at any time within the automobile. The second proposal would be to reduce the expansion power of the airbag by approximately 25%; the resulting lower force would be safer for smaller individuals.

What are **smart air bags**?

Several major automakers are now involved in making safer air bags, striving to become the safest manufacturers in the market. One way of making air bag deployment safer would be to adjust the speed of expansion for the air bag itself. Smart air bag deployment would rely on sensors located throughout the seat to determine the seating position, height, and weight of the passenger. Although manufacturers have not placed smart sensors in any vehicles yet, they should be quite popular within the next few years.

Front air bags are quite common, but what are **side air bags**?

As an optional safety feature in 1995 model vehicles, Volvo installed side air bags that deploy during certain side-impact collisions. The position

Air bags on Mars?

Probably the most expensive air bags ever created, the NASA Pathfinder's air bag cost approximately $5 million dollars to develop and install. Why would the Pathfinder need an airbag on Mars? Although it did not collide with high-speed Martian vehicles, it did collide with the planet's surface when landing. The gravitational pull of Mars accelerated the probe toward the planet's surface at great speeds. A parachute, along with retro-rockets, slowed down the falling probe, but four 8-foot radius air bags were needed to cushion the 65-mph landing. The Pathfinder survived by increasing the time in the collision, and in turn decreasing the force felt by the space probe.

of the side air bag is vital, for it prevents the head and upper body of the occupant from striking the metal or the glass in the door. This, along with the side-impact reinforcing bars within the doors, have the potential to save as many lives as air bags and crumple zones have in head-on collisions.

Side-impact air bags are just one of several reasons why Volvo has a reputation for safety "firsts." The most famous of its safety features was the three-point seat belt. Other Volvo "firsts" include the first frontal headrest, the first laminated windshield, and the first side-impact bar.

Conservation of Momentum

What is the **conservation of momentum**?

According to the conservation of momentum, there is a fixed amount of momentum for the entire universe. Additional momentum cannot be gained or lost in the universe, but only transferred from one object to

Why can't space ships use propellers or conventional airplane jets to move in space?

A rocket cannot maneuver in space with propellers or airplane jets because those machines need to push against the air in order to accelerate, and there *is* no air in space. Rockets do not need to push against air because they rely on the action and reaction of the exploding fuel inside the ship itself. As the exhaust accelerates out of the rocket, the rocket in turn accelerates forward.

another. For example, if an ice skater with a mass of 50 kg (kilograms) was skating with a velocity of 2 m/s (meters per second), his momentum would be 100 newton-seconds. If another person were at rest on the ice, that person would have a momentum of 0 newton-seconds. Both the skaters together would have a total momentum 100 newton-seconds. That amount of momentum will be conserved regardless of any collision between the skaters.

The 100 newton-seconds of momentum can be transferred between the skaters. If the moving skater collides with the stationary skater, some of the momentum of the moving skater would be transferred to the stationary skater. Depending upon the type of collision, the momentum of each skater would be different. The moving skater would lose momentum because it would transfer some of its momentum to the other skater, yet the momentum of skater #1 plus skater #2 would remain at 100 newton-seconds.

ROCKETS

How do rockets use the **conservation of momentum**?

Launching a rocket is an example of recoil and the conservation of momentum. Upon launching, rockets burn fuel and expel exhaust.

Because the rocket is pushing out the exhaust at a high velocity, the exhaust has momentum. Newton's Third Law and the conservation of momentum state that the rocket must have an equal and opposite force and momentum. Since the rocket has more mass than the fuel's exhaust, the rocket moves upward at a slower velocity than the exhaust's velocity.

The rocket accelerates upward and eventually attains a huge velocity for two reasons. First, as the firing continues, the amount of fuel decreases in the rocket, reducing its mass and allowing it to accelerate. Secondly, since the burning fuel also creates a constant force, the rocket builds up more and more speed as the rocket's engines burn. Once the rocket leaves the Earth's atmosphere, it encounters no air resistance and can shut down the rockets and travel at a constant velocity.

How do rockets **maneuver** in space?

Because there is no air in space, wings and flaps would do little to help steer a rocket. The space shuttle, for example, uses this equipment only upon re-entry into the Earth's atmosphere. In outer space, the shuttle fires different rockets—called thrusters—that work just like conventional rockets. The thruster is fired in the opposite direction in which the shuttle wants to move. If the shuttle wanted to slow down, a thruster would be fired forward, giving the shuttle momentum and acceleration backwards.

RECOIL

What causes recoil when firing a **gun**?

The explosion of gunpowder creates a large force that accelerates the bullet out of the barrel. The conservation of momentum declares that the momentum of the bullet must be equal and opposite to the momentum of the gun. Although the amount of momentum is the same for both the gun and the bullet, the large mass of the gun causes it to move slower than the bullet. And although it moves slower, it does have enough momentum to injure the person firing the gun.

How can a person **reduce the risk** of getting hurt from the recoil of a rifle?

The key to avoiding injury from recoil is to hold the rifle tightly against one's shoulder. By attaching the rifle to the shoulder, the person's mass becomes part of the gun-person system. The conservation of momentum states that the momentum of the gun must be equal to the bullet. Since the combined mass of the gun and person is greater than the gun by itself, the gun's velocity in the recoil would be much less.

PROJECTILE MOTION

What is a **vector** and what is it used for?

A vector is a quantity that has magnitude (a numerical measurement) and direction. Velocity, for example, is a vector, because unlike speed (which is not a vector, but a scalar), velocity comprises a direction in addition to a rate of speed, such as 40 mph (miles per hour) north.

When describing physical occurrences or trying to solve problems in physics, it is often easier to describe the problem by drawing a picture. If a variable in the problem deals with motion, a vector is used to describe that motion by drawing an arrow. The length of the arrow indicates the magnitude, while the tip of the arrow shows the direction of the vector quantity.

For example, if a car is traveling with a velocity of 55 mph east, then a vector could be used to describe that motion. The length of the vector would represent the 55 mph speed, while the arrow would point in an easterly direction. Vectors are used throughout physics to describe various types of physical motion and forces.

Have vectors always been used in **physics**?

Although there have been methods to describe physical quantities that are close to the modern vector for hundreds of years, it was not until about a

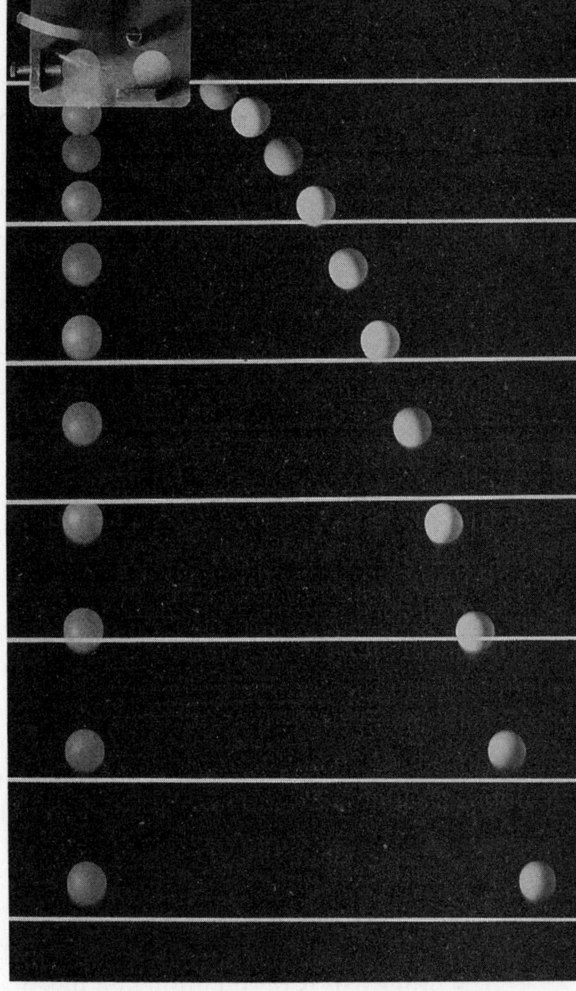

A ball dropped from a given height will hit the ground at the same time as one projected horizontally from the same height.

century ago that a British mathematician developed the vector as we know it today. Oliver Heavyside helped simplify and develop the concept of today's vectors to represent motion. His work in vectors has proven to be extremely helpful in solving both simple and complex physics problems.

Which object hits the ground first: a bullet fired **horizontally** or a bullet **dropped** from the same height?

This is a famous physics question posed to introductory physics students. The typical response would be the bullet that is dropped lands

When is the best time to drop a bomb... before, over, or after passing the target?

The best time to drop a bomb would be a pre-determined distance *before* the target. Although the bomb is falling, it is still traveling forward with the same velocity as the aircraft that was carrying it, according to the Law of Inertia. Therefore, to have the bomb land on target, the bomb must be dropped before the target.

first, because it has less distance to travel. Although this seems to make sense, the answer is wrong. In fact, both objects land at the same time.

All objects are pulled toward the Earth by gravity, and accelerate at a rate of 9.8 m/s^2 (meters per second squared). This means that all objects fall at the same rate. If a bullet is given a horizontal velocity in addition to its downward acceleration due to gravity, it will continue to move across as it falls down—but it still keeps falling at 9.8m/s^2, just like the bullet that is dropped. Both the dropped bullet and the bullet fired horizontally fall at the same time; the added bonus for the fired bullet is that it travels horizontally as well.

The strobe photograph on the previous page shows one ball falling while the other ball is both falling and given a horizontal velocity. Notice that at each time interval, both balls have fallen the same distance.

What angle does a **cannon** need to attain the **greatest range**?

In the absence of air resistance, a cannon should be fired at an angle of 45°. This angle produces the greatest range because it lies directly between a horizontal path, which will hit the ground just after it is fired, and a vertical path, where the cannon ball would rise straight up and then fall directly downward. The horizontal component will give the cannon ball enough sideways motion, while the vertical component of the launch will give the cannon ball enough height to stay in the air a little while.

How does **air resistance** affect the path of a cannon ball?

Air resistance acts as a frictional force against the motion of the cannon ball. If normal air resistance—with no wind—were taken into consideration, the best angle for greatest range would be approximately 35° up from the horizontal.

ORBITS

How does an object **stay in orbit** around the Earth?

A cannonball, as Newton suggested, could orbit the Earth if given enough horizontal or sideways velocity. All objects continually fall toward the Earth, yet if that falling motion is combined with a large horizontal motion, by the time the cannonball were to hit the ground, the surface of the Earth would have already curved away from the cannonball; therefore, the cannonball continues around the Earth, establishing an orbit. In effect, the cannonball—or any satellite—would continually fall toward the Earth, yet continually miss it as well.

Does the **space shuttle** orbit around the Earth or fall toward it?

The space shuttle falls toward the Earth just as the cannonball falls toward the Earth in the previous question, with one important exception. The shuttle, like the cannonball, has a large enough horizontal velocity along with its free fall. Although the shuttle falls toward the Earth, it misses it because the horizontal velocity of the shuttle allows the Earth's surface to curve underneath it before the shuttle can crash into the Earth. Falling to the Earth while never striking it is called orbiting.

Are the astronauts truly **weightless** when they are orbiting the Earth?

Whenever someone jumps off, say, a ladder, and is in free fall, that person becomes weightless until he or she hits the ground. Gravity is still

Astronaut Bruce McCandless floating near the Space Shuttle Challenger.

acting on astronauts while in free fall, but since there isn't a floor or supporting structure to hold them up, the astronauts feel no weight. Therefore, feeling no weight, even while in free fall, means they do not weigh anything. If you have a difficult time with this concept, imagine yourself falling. Now place a scale under your feet while you are falling. The scale will read zero.

How fast would a **baseball** need to be hit in order for it to be **"put into orbit"**?

It is not feasible to put a baseball in orbit because air resistance, tall buildings, and mountains would get in the way. However, if one did not have to consider air resistance, physical obstructions, and the fact that there is no one who can hit or throw a baseball at such a high speed, the ball would have to travel at a velocity of 7.9 km/s (kilometers per second) or 17,800 mph (miles per hour). At that rate, the baseball would orbit the Earth in approximately 84 minutes.

How high above **sea level** does the **space shuttle** orbit the Earth?

In order for the space shuttle to efficiently orbit the Earth, it must avoid the air resistant forces of the Earth's atmosphere. Therefore, the space shuttle (and most satellites) orbits the Earth approximately 200 kilometers above sea level. At this altitude, the space shuttle takes about 1.5 hours to orbit the Earth. It is very difficult to alter the time it would take the shuttle to orbit the Earth, since it is relying only on gravity. If the shuttle orbited slower, it might not have enough horizontal velocity and crash into the Earth. On the other hand, if the shuttle were given a greater velocity, it would take on an elliptical orbit. And if the velocity were even greater, it could achieve a parabolic path and leave the Earth for interplanetary space.

What is **escape velocity**?

In order to leave the Earth and travel into interplanetary space, a space probe would have to achieve a speed of 11.2 km/s (kilometers per second), or 25,000 miles per hour, for a brief period of time. This velocity, called the escape velocity, is the velocity that any object would need to escape the Earth's orbit. If a space probe were to achieve such a speed while in orbit, the probe would achieve a parabolic path—swinging in rather close to the Earth's atmosphere and whipping around like a rock in a sling shot, having enough energy to overcome the gravitational force of the Earth.

What are some **escape velocities** that objects much achieve to "escape" from the other planets in our solar system?

Planet	Escape Speed (km/s)
Mercury	4.3
Mars	5.0
Earth	11.2
Earth's Moon	2.4
Venus	10.4
Jupiter	60.2
Saturn	36.0

Planet	Escape Speed (km/s)
Uranus	22.3
Neptune	24.9
Pluto	not available

What was the **first space probe to leave** our solar system?

The first space probe to leave our solar system was *Pioneer 10.* Launched by NASA in 1972, *Pioneer 10* was to travel past and observe the outermost planets of our solar system. After passing the orbits of Pluto and Neptune, *Pioneer,* along with the later space probes, will travel aimlessly into space, becoming the first human-made devices to leave our solar system.

How does an **interplanetary probe** navigate through our solar system?

It would be impossible for a space probe to carry enough fuel to fire its way out of the solar system. Instead, a probe travels through space, relying on its inertia and gravitational forces from planets. To power their computers and navigational systems, the probes operate not on solar power, but on electricity generated by the heat from decaying radioisotopes. The probe will also use the potential and kinetic energy it gains from the planets' gravitational forces to accelerate it on its way out of the solar system.

CIRCULAR MOTION

What is the difference between a **revolution and a rotation**?

Although the two terms are often used interchangeably in everyday conversation, they have two different meanings. To rotate is to spin about an axis located within the spinning object. An example of a rotating

Which travels faster—the inside or the outside of a record album?

There are two ways to approach this question. One of them is to say that both the inside and the outside of the record travel at the same speed. Both points start and stop at the same point and take the same amount of time to make one revolution. This measurement is called angular speed. The angular or rotational speed is measured by counting the number of revolutions the record makes in a particular amount of time.

The other approach is to measure linear speed, the distance traveled over a period of time. The geometry of circles states that the distance on the outside of a circle has a greater circumference than a point closer to the center of the circle. Therefore, the outer point of the record travels faster than the inside, for it travels a greater distance over the same period of time as the inside position.

object is the spinning Earth. The Earth rotates about its inner north-south axis every twenty-four hours. A revolution, however, takes place when an object turns about an axis that is located outside the object. The Earth also revolves about the sun, its outer axis, every 365 days. The Earth both rotates and revolves, but each term describes a specific type of turning motion.

What is a **centripetal** force?

A centripetal or "center seeking" force causes objects to travel in circular paths. All objects that curve or turn experience this force. The force that is responsible for twirling a set of keys from a string is the centripetal force from the string. This force is at a right angle to the motion of the keys, and if there were no centripetal force acting in toward the center of the circle through the tension in the string, the keys would continue in a straight line and not revolve about the person's hand.

The formula for centripetal force is $F = mv^2 / r$, that is, centripetal force is equal to mass times velocity squared divided by the radius.

How come people don't fall out of **upside-down roller coasters**?

Whenever there is a rotation or revolution of some type, there is a centripetal force. For example, the Earth applies a gravitational centripetal force on the moon. It is this centripetal force, along with the moon's tangential velocity, that causes it to orbit the Earth in a nearly circular path.

Roller coaster riders experience centripetal forces when going through loops and spinning inside rides. The centripetal force applied to them is typically exerted by the track or wall of the ride, in that the track or wall prevents the riders from flying off or out of the ride. Again, the force is applied toward the center of the circle, but the object's velocity is directed in a straight-line tangent to the circle. This concept is the reason upside-down roller coaster loops are possible.

What is a **centrifugal force**?

Centrifugal force is not really a force; it is the sensation that a person feels when traveling in a circular path. When experiencing a "centrifugal force," a person feels as though they are being pushed out, away from the center of the circle, but that is not the case. The person's inertia wants them to travel in a straight line, tangent to the circular path, but the centripetal force pushes them back in toward the center of the circle, keeping the person from traveling away, in a straight line from the circle. Centrifugal force is a simulated force.

Why are some **turns** on an **incline**?

Banked roadways are often found on highway exit ramps and on racetracks, where large velocities can make turning very dangerous. A banked roadway is a road that is on a slight tilt toward the center of the turn.

The law of inertia says if a vehicle is traveling at a fast speed, that vehicle wants to continue traveling at that speed in a straight line. If a high-velocity vehicle attempts to turn on a horizontal roadway, the cen-

Upside-down loop of a roller coaster.

Banked roadway at the Daytona International Speedway helps keep the cars on the road (see page 68).

tripetal force responsible for turning the vehicle is the frictional force between the tires and the road. However, if an incline is introduced, the centripetal force relies not only on the frictional force between the tires and the road, but also on the normal or supporting force from the road; these direct the car toward the center of the circle of the turn, preventing the car from flying off the road without turning.

How can someone be suspended against the wall inside a **rotating amusement park ride**?

There are three forces responsible for keeping the person suspended against the wall of this type of amusement park ride. The primary force is the centripetal force applied by the wall of the rotor machine; this wall prevents the participant from traveling in a straight line, tangent to the circular path. When the rotational speed of the ride is sufficient, the centripetal force applied by the wall is large enough for the person to feel as though he or she is being pushed outward (the fictitious centrifugal force). The simulated gravity is caused by a person's inertia wanting to take them tangent to the circular path.

The second reason for the suspension is the frictional force between the wall's surface and the person's clothing; this force helps prevent the person from sliding down the wall. Finally, the person's weight is the last of the three major forces. This force points downward and cancels out the upward frictional force, leaving the rider suspended against the rotating wall.

Amusement park ride dependent upon centripetal force.

How can **water,** when spun around in a bucket, remain inside the bucket?

If a bucket of water were spun around in a vertical circle (such as over a person's head), it would seem that the water should fall out due to the gravitational force acting on the water. However, if the rotational speed were fast enough, the water in the bucket has a horizontal velocity and downward fall that is the same as the bucket being swung around. Because the bucket doesn't fall straight down, neither does the water. This example is similar to why satellites orbit the Earth and how we remain in our seats in a roller coaster loop.

Would the **centripetal force** be larger for a car making a **sharp turn** or a more **gradual turn**?

If the velocities were the same, the amount of force experienced on a sharp turn would be greater than the force that is experienced on a gradual turn. For example, two cars travel at the same linear speed and are rounding a turn. The car on the inside needs to apply more force on the steering wheel and tires to turn the car, whereas the vehicle on the

Artist's impression of the inside of a rotating space station, which would simulate gravity through centripetal force.

outside has a more gradual turn, thus it needs to apply less centripetal force in order to turn. As the radius decreases, the centripetal force required to turn increases.

Why does an airplane **bank into or toward** the turn?

In order to execute a turn in the air, a component of the lifting force of the airplane (which keeps the plane in the air and usually acts perfectly upward) must point toward the center of the circle. This component of the lifting force is the centripetal force that causes the aircraft to turn. Once the plane has completed its turn, the plane evens out into its normal horizontal position with the lifting force acting upwards. (For more information on lifting forces and flight, refer to the FLUIDS chapter.)

How might circular motion and centripetal force play a part in **future space stations**?

A circular space station would rotate at a predetermined velocity to create a simulated gravitational field. Simulated, because a person or astronaut inside the rotating space station would want to travel in a line tangent to the circular motion of the space station, yet the centripetal force from the walls of the space station would prevent this escape from occurring. As a result of the centripetal force and the person's inertia, the astronaut would feel as though he or she were being pushed out against the wall of the space station. The wall in this case would act like the ground, and the astronauts would be able to walk around on the walls—as if they were the ground—because that is where their bodies want to move toward. The round shape of the space station would allow a person to walk "inside" the circular walls of the station, much as we, on the Earth, walk on the "outside" of the circular surface of the Earth. Finally, if the radius and rotational speed were correct, the astronaut could perhaps experience a "simulated" acceleration due to "simulated" gravity of 9.8 m/s^2.

Artist's impression offering a view of the outside of a rotating space station.

ROTATIONAL MOTION

TORQUE

What is **torque**?

Torque is a force that, when applied to an object, causes a turning motion. Torque is used to open doors, tighten screws, rotate wheels, and ride seesaws. Although forces are used on all these objects, these forces do not cause the objects to accelerate in a straight line, but instead to turn.

Torque is calculated by multiplying the force that is applied to the object by the distance the force was from the axis of rotation. For example, if a force of 50 newtons is applied to a door (by pushing on the doorknob) and that force is applied 1 meter away from the axis of rotation (the hinges), the total amount of torque used to open the door would be 50 newton-meters.

What is a **lever-arm**?

A lever-arm is the distance between the axis of rotation and the point where the force is applied. In the above example, the lever-arm would be the distance between the hinges (axis of rotation) and the doorknob (force point). Whenever torque is applied to an object the result is leverage, and leverage or torque is the force multiplied by the length of the lever-arm.

Why is it easier to loosen a **nut** with a **longer wrench**?

A longer wrench has a longer lever-arm, and a longer lever-arm provides more torque to help turn the nut. The added torque means that less force is needed to loosen the nut than would be necessary if using a shorter wrench.

Why are **doorknobs** typically as far away as possible from the **hinges**?

The reason why doorknobs are located as far away from the hinges as possible is so that a person does not have to apply as much force to open the door. For example, suppose it takes 50 newtons to open a door that is one meter wide. If the doorknob were in the middle of the door (0.5 meters away from the hinges) then the person would need to apply 100 newtons of force on the doorknob to open the door. However, if the doorknob were at the outer edge of the door (where doorknobs typically are), then the person would only need to apply 50 newtons of force to open the door. In both situations the torque was 50 newton-meters, but the force that the person had to exert was half as much when the doorknob was 1 meter away from the axis.

ROTATIONAL INERTIA

How is **rotational inertia** different from inertia?

Inertia is an object's resistance to changing its motion. Rotational inertia, otherwise referred to as the moment of inertia, works on the same principle. Just as with regular inertia (an object at rest will always stay at rest or a moving object will continue the same motion until another force causes the object to change its motion), a rotating object will continue to turn until a torque causes the object to change its rotational motion. Rotational inertia, or the moment of inertia, is the same as linear inertia, except for the fact that rotational incorporates a turning motion.

Skater Michelle Kwan tucks in her arms to execute a spin.

How is rotational inertia **measured**?

Just as inertia is determined by how much mass an object has, rotational inertia is determined not only by the amount of mass, but where the mass is located in reference to the axis of rotation. Rotational inertia increases with the square of the distance between the mass from the axis of rotation.

Why do **ice skaters** spin faster when they tuck in their arms and legs?

If an ice skater extends both arms and one leg while spinning, he has moved a rather large amount of mass away from his axis of rotation, which increases the rotational inertia, causing the skater to spin rather slowly. If the arms and legs are in a "tucked" position, then the mass is much closer to the axis of rotation, and the rotational inertia is decreased, allowing the skater to spin much faster.

ANGULAR MOMENTUM

What is **angular momentum**?

Just as linear momentum (straight-line momentum) is the inertia of an object multiplied by its velocity, angular momentum is the rotational inertia of an object multiplied by its rotational velocity. A conservation law holds for angular momentum as well, and states that angular momentum cannot be gained or lost, only transferred from one rotating object to the next.

Why do **cats** always land on their **feet**?

Cats are notorious for landing on their feet when dropped from an upside-down position. The reason for this involves rotational motion and angular momentum, even if the cat is not initially spinning. Many people incorrectly assume the tail causes a cat to rotate when falling, yet cats without tails can spin and land on their feet just as easily as cats with tails.

When dropped from an upside-down position (with no initial rotational movement), the cat has an angular momentum of zero and keeps that net angular momentum of zero throughout the entire fall. In the first section of the fall, the cat will extend its hind legs and tuck in its front legs. Because of this, the hind section of the cat would have a relatively large rotational inertia compared to the front section, allowing the front section to rotate faster in a counterclockwise direction, while the back section rotates very little in the clockwise direction, keeping a net angular momentum of zero. When the front of the cat's body is in position, the same maneuver is performed, but this time with the front legs extended and hind legs tucked in. The cat's rear section would have less rotational inertia and spins more than the front. Finally, the cat is in position to land on all fours, while keeping a constant angular momentum of zero.

GYROSCOPES

What is a **gyroscope**?

A gyroscope is typically a spherical or disk-shaped object that is free to rotate in any direction with very little friction. Gyroscopes are often used to illustrate the law of rotational inertia, which states that a rotating object will continue to rotate until an outside torque changes its

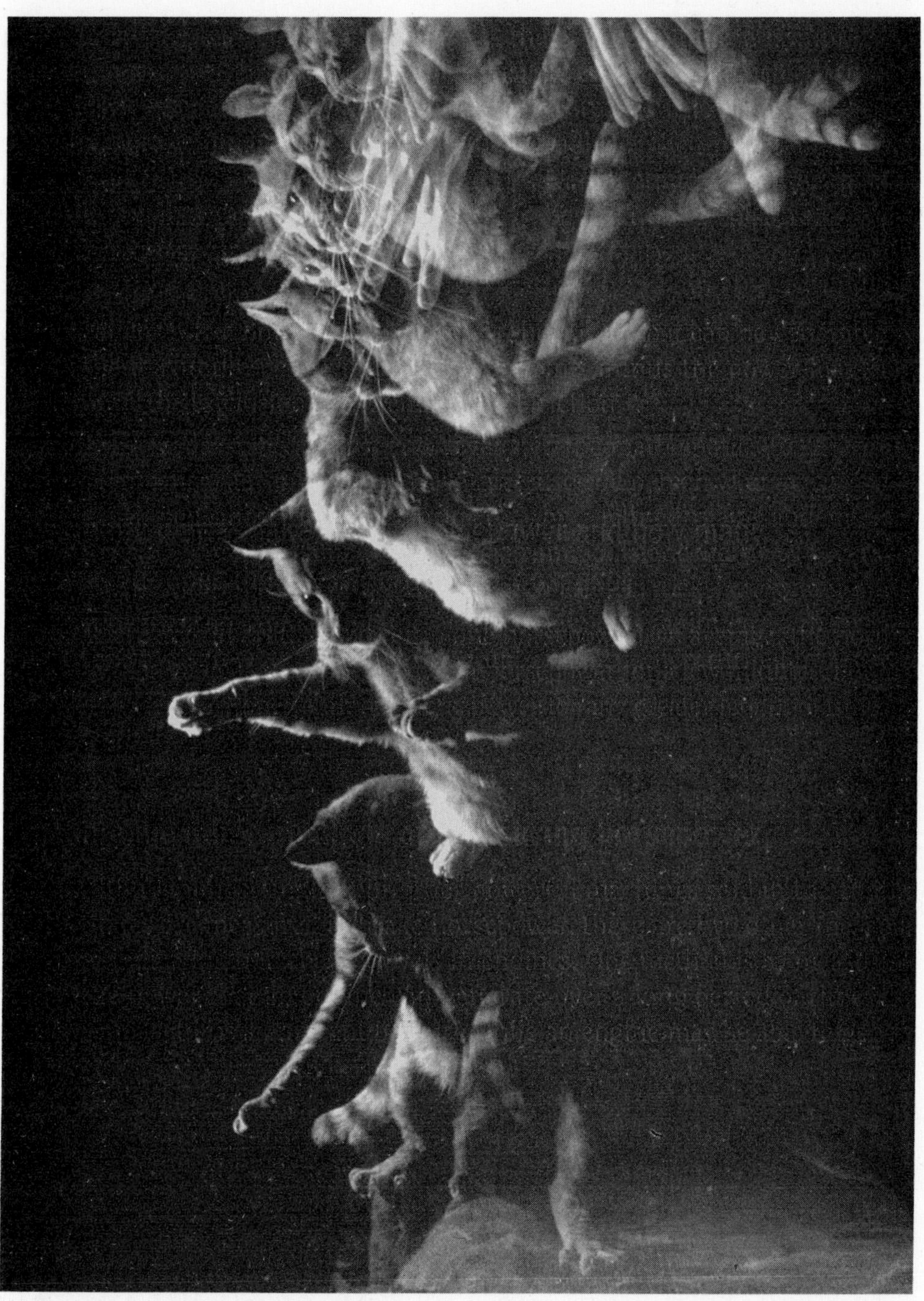

A strobe photo of a falling cat landing on its feet.

A gyroscope.

rotation. The Earth is a wonderful example of a gyroscope; it continues to rotate about its axis and will forever continue to do so until some outside torque changes that motion. Famous with children for years, tops are another example of gyroscopes. When spun, the top will continue to rotate in a vertical orientation until the friction between its pointed end and the surface creates too much torque and causes the top to precess; this means it no longer spins directly vertical, but tilts a bit while rotating, and continues to do this until the friction slows down the rotational velocity and the top strikes the floor.

What are some **examples of gyroscopic motion**?

The rotating Earth and a toy top are typical examples of gyroscopes. Other everyday applications of gyroscopic motion are all around us as well. For example, a football is much easier to throw when it acts as a gyroscope. When thrown with a spiral, the football keeps the same orientation for the entire duration of the throw. If the nose of the football is tilted up when thrown with a spiral, the nose will still be tilted up when caught. A bullet fired out of a muzzle of a gun rotates as well; by rotating (and therefore gaining gyroscopic characteristics), the bullet cuts through the air and maintains its desired trajectory instead of flying end over end to the target.

What are **gyroscopes used for**?

Along with bullets, footballs, tops, and the Earth, there are many other uses for gyroscopes. Gyroscopes called gyrocompasses play an important role in navigational and guidance systems in planes, ships, rockets, and missiles. Gyrocompasses read true north, as opposed to magnetic north; this reliance on rotational inertia and torque means that they do not produce erroneous readings as magnetic compasses can produce when placed near electrical equipment. By detecting any changes from a set course, the gyrocompass can even send out signals to a navigational system, and at times is used to help stabilize boats in heavy seas by measuring the defection of the ship compared to the gyrocompass.

How do **automatic pilot systems** employ gyroscopes?

Autopilot navigation systems on airplanes use not one, but usually a number of gyroscopes, to assist navigational systems in determining where they are and where they have to go. A vertically oriented gyroscope detects changes in pitch (nose up/nose down orientations) and roll (side-to-side tilting) by creating an artificial horizon. The artificial horizon is a vertical line, representing the horizon that the plane's navigational systems measures itself against. Another set of gyroscopes determine the plane's direction or heading. This directional gyroscope is similar to the gyrocompass used on many ships. The computer that controls the automatic pilot settings determines how to react to the different readings and makes adjustments accordingly.

How does a bike stay **upright**?

Any bike-riding child can tell you that it is easier to remain upright on a moving bike. Although the child might not be able to tell you why this is the case, physicists know that it has something to do with rotational inertia and gyroscopic motion of the wheels. A bicycle tire acts like a gyroscope, because it rotates around a relatively low-friction axis, and will continue to do so until friction slows it down too much. When moving, a large tire has a great deal of rotational inertia, which makes it more difficult to tip when an external toque is applied. If the bike's frame leans left, the front wheel will automatically steer itself to the left in an attempt to maintain an upright position (for if the wheel did not turn, the bike and rider would fall). By tilting the bike when turning we are able to remain stable when making sharp turns.

WORK

What is **work**?

In everyday language, doing work means you've expended energy in order to accomplish something. In physics, that expended energy is a force applied to an object in the direction the object is to be moved, and the accomplishment is that the object moves a distance. A simple mathematical formula defines work as force × distance.

What is a **joule**?

A joule, named after James Prescott Joule, a nineteenth-century British physicist, is the conventional unit for work and energy. A joule is a metric unit for work and is defined as a force of one newton acting over a distance of one meter, where the force and the distance moved lie in the same direction. Other units of work are the erg, foot-pound, and the Btu, or British thermal unit.

What are some significant **conversion factors** for work?

A joule is the standard unit of measurement for work and energy in the metric system. Other units, however, can be used in other systems of measurements. The following are conversions from joules to foot-pounds, ergs, and Btus:

1 joule = 0.7376 ft-lb

1 joule = 1×10^7 ergs

1 joule = 9.4×10^{-4} Btu

What is the **difference** between **work and energy**?

There actually isn't much of a difference between work and energy. In order to do work, an object must have energy, and in order to have energy, an object must have work done to it.

POWER

If something is **moved quickly,** does that **increase the amount of work** done?

If a 10-newton force is applied to an object, moving it 10 meters in 10 seconds, the person has done 100 joules of work (force × distance = work). If the person does the work twice as fast, the amount of work is the same (100 joules), but the person has generated twice as much *power.* The formula for power is the work divided by time, or how fast the work was done. The unit for power is the watt, named after James Watt.

The power of engines in cars, motorcycles, and lawnmowers is **measured in horsepower.** What does that mean?

A consumer who wishes to purchase a machine that has an engine typically wants to know if it will have enough power to get the job done. Most engines in the United States are measured by their displacement, while their power output is measured in horsepower. However, there are three different ways to measure horsepower that all produce unique results. The United States standard method of measuring horsepower is

Where did the term horsepower originate?

The term horsepower came from Scottish inventor James Watt. The value for a unit of horsepower was determined after Watt performed several experiments on horses pulling coal. He originally determined that the average horse was able to pull 22,000 foot-pounds every one minute. In other words, a horsepower was defined as the amount of power exerted to move 22,000 pounds of coal by one foot in one minute. Watt was not happy with his figure because he felt it was too low; he thought the average horse was more powerful than his original calculations and experiment indicated, so after extensive study of horses, he increased the value of a horsepower to 33,000 foot-pounds per minute.

called "brake horsepower," which is the power that can be generated at the maximum performance of the engine, minus the power lost from heat, the expansion of the engine, and friction.

How many **watts** make up one horsepower and how many horsepower does the **average person possess**?

There are 746 watts of power in one horsepower. On average, the normal human can generate one-third of a horsepower for a few minutes, but can only generate one-tenth of a horsepower for any greater length of time.

What are the **other methods of measuring horsepower**?

Although most engines or other powerful machines sold in the United States are measured in brake horsepower, there are two other methods of indicating horsepower levels in engines. One method is by measuring the power that an engine can generate over a long period of time; the time factor is significant because while an engine may be able to generate a great deal of power, it might not be able to sustain that level of production for any length of time.

The other method is simply indicating what the ideal power output would be from an engine, without regard for the energy converted to heat, which would not be a usable form of energy for the engine.

Conservation of Energy

What kind of **energy** do you **gain riding up** in an elevator?

As an elevator's motor does work on the elevator and its passengers, it is giving them energy. Specifically, when an elevator elevates itself and the people it contains, it is providing gravitational potential energy. The energy is potential because the elevator and the people have the potential to move should the elevator cable break. This type of potential energy is gravitational because the elevator and the unlucky riders would accelerate downward toward the Earth due to the gravitational attraction between the riders and the Earth.

What kind of **energy** do you **gain when falling** to your death in an elevator?

Potential energy is dependent upon the weight of the object and the distance it is from the ground. The higher up an object is, the greater its potential energy. When an object descends, however, the gravitational potential energy is transferred to moving, kinetic energy. So whether an elevator is plunging people to their deaths or gently lowering them from Housewares to Ladies Undergarments, the elevator accelerates, gaining kinetic energy while losing its height and gravitational potential energy. Kinetic energy (small reward for a death plunge) is defined as one-half the mass times the velocity squared ($m/2 \times v^2$).

How does the **conservation of energy** apply to a falling elevator?

In the previous example of a falling elevator, the energy of the elevator and its passengers changed gradually from potential energy to kinetic energy. Although the energies changed form throughout the elevator's fall, the total energy of the elevator and its passengers is fixed, potential plus kinetic; as the elevator falls, the velocity and the kinetic energy increase, while the potential energy decreases.

What is the **conservation of energy law**?

Throughout the entire universe, there is a fixed amount of energy and it will never change. The law specifically states that energy cannot be created or destroyed; it can only be changed from one form of energy to another. In other words, there is a fixed amount of energy, so we really can't use it up. When we use energy, we are actually changing it to another form. The total amount of energy in the universe does not change.

The conservation laws—conservation of energy, conservation of linear and angular momentum—are the very basis of most of what is "modern physics," that is, Relativity, quantum mechanics, and the like.

SIMPLE MACHINES

How can we make work **easier**?

Machines do not reduce the amount of work for us, but they can make it easier. There are four types of simple machines known in engineering and physics; these were discovered long ago, and still form the basis for all mechanical machines today.

Work is force multiplied by distance. Machines are intended to reduce the amount of force that needs to be exerted, but in the process, this increases the distance that the force must move the object.

What is **mechanical advantage**?

Mechanical advantage is an indication of how much the effort force is reduced by a machine. If a machine allows you to exert one-half the

Are machines with no mechanical advantage still useful?

Machines with a mechanical advantage of 1, that is, the machine provides no reduction either in force or in distance, are still useful because they can change the direction between the force and the movement itself. For example, a single pulley does not reduce the amount of "pull" needed to lift an object, or the distance the object needs to be lifted, but it does allow one to pull up an object from below, rather than having to push it.

force normally needed to do work on an object, the mechanical advantage is 2. If the machine allows for one-third the force, the MA is 3. A machine with no net mechanical advantage—that is, a machine that does not reduce the force or distance—has a MA of 1. If a machine has a mechanical advantage of less than 1, it increases the force needed but decreases the distance that it needs to be moved.

Why would you want a machine that **increases the force needed**?

Machines with a mechanical advantage smaller than 1 are useful, for although they increase the force needed to perform work, the effort force only needs to travel a very short distance; therefore, they work faster than machines with greater mechanical advantages.

What are the **four types of simple machines**?

1. Inclined plane
2. Lever
3. Wheel and axle
4. Pulley

Movers using a ramp, or inclined plane, to make their job easier.

INCLINED PLANES

How is a **ramp** an example of an inclined plane?

Inclined planes are most useful when attempting to lift something that is very heavy. For example, movers use inclined planes instead of lifting heavy objects directly into trucks. By using a ramp (an inclined plane), movers can drag the object up the incline; this requires less force, although the distance up the ramp is greater than the distance straight up into the truck. Although less force is exerted because the distance is greater, the work done is the same as if the object were lifted straight up into the truck.

Who first used the **inclined plane as a simple machine**?

No one knows for sure who invented the inclined plane, but it is believed that the Egyptians used inclines when building the pyramids. Some historians feel the inclines were over a kilometer long so the slaves could drag and push the massive blocks of masonry, which weighed several hundred tons. Although the distance traveled was greater, the effort force was dramatically reduced, making it seem easier for the workers.

Interestingly, other historians hold that raising the stones of the pyramid was done by levering; in fact, ancient paintings of this process have been found in a number of locations in tombs and on large monuments in Egypt. These historians reject the ramp theory because they believe that at some point, more effort would have gone into building the ramps than building the pyramids the themselves. For instance, the great pyramid of Cheops would have required a ramp with many times more material in it than the pyramid itself. The problem with an earthwork ramp is that the volume of material needed to keep the ramp stable is approximately the height of the ramp to the third power.

Is a **wedge** a simple machine?

A wedge is an inclined plane that can be moved. Chisels, knives, hatches, carpenter's planes, and axes are all examples of a wedge. Wedges can have only one sloping plane, as in a carpenter's plane, or they can have two, as in a knife blade.

What type of simple machine is a **screw**?

A screwdriver is a tool (see under the "Wheels and Axles" heading), but so is a screw itself. If one could unwrap the twisted edge of a screw from its shaft, a long inclined plane would emerge.

Screws can be used in two major ways. First, they can be used to hold things together. Some simple examples include wood and metal screws and the screws on jars and bottles and their tops. Screws can also be used to apply force on objects. The screws found in vises, presses, clamps, monkey wrenches, brace and bits, and corkscrews are some examples of this application.

The screw acts as a simple machine when an effort force is applied to the larger circumference of the screw. For example, a person might apply the effort force to a wood screw by turning a screwdriver. That force is

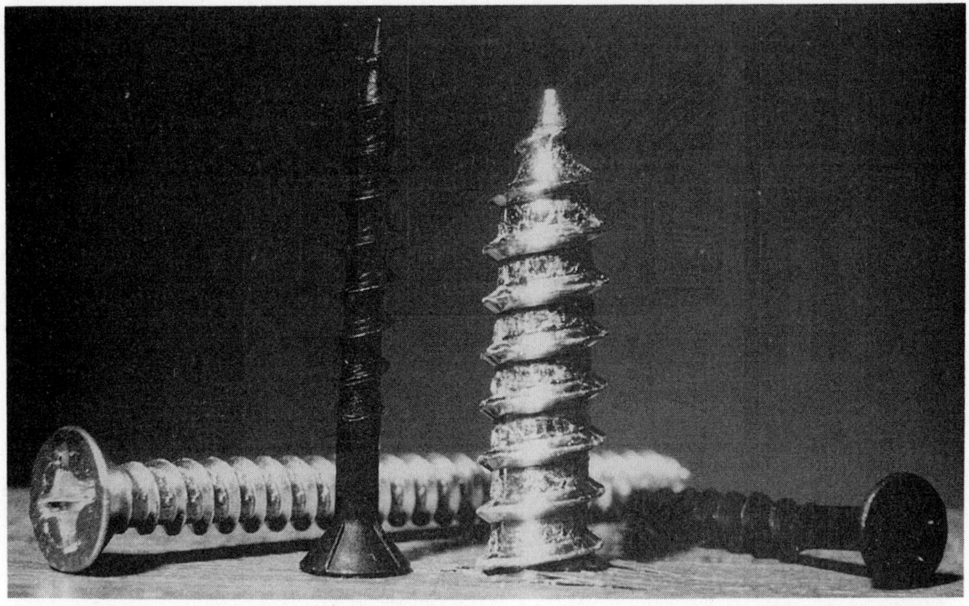

Various types of screws, all made from twisted inclined planes.

then transmitted down the spiral part of the screw (the thread) to the tip of the screw. The movement of the screw tip into the wood is the resistance force in this machine. Each complete turn of the screwdriver produces a movement of only one thread of the screw tip into the wood. This distance between two adjacent threads is called the pitch. The lower the pitch of the screw (that is, the closer the threads are to each other), the greater the mechanical advantage and the easier it is to screw into the wood, because there are more turns of the screw—i.e., a greater distance—resulting in less force necessary.

Who **invented the screw**?

Archimedes, who studied and developed the mathematics to determine mechanical advantage for levers, was also the person who invented the Archimedean screw (a variation of the inclined plane and modern screw), which was able to lift water from pools and wells. By reversing the motion of the screw, material such as dirt, rocks, and water could move up the ramp or incline of the screw. This is similar to the way in which a screw or drill bit brings up saw dust when turned in reverse.

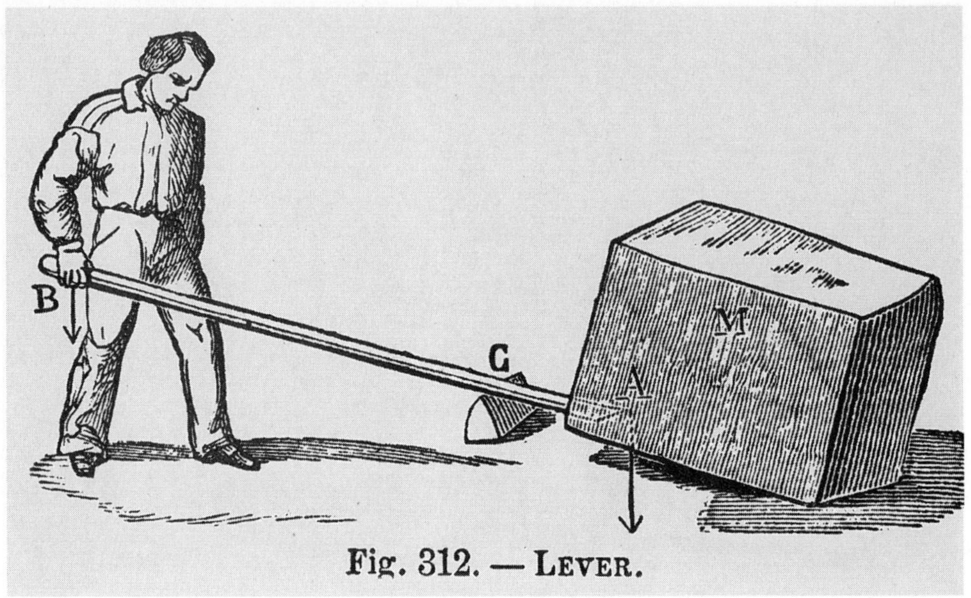

A lever and fulcrum.

Each rotation of the screw would lift material up the distance between two threads on the screw.

LEVERS

What is a **lever**?

A lever is a simple machine that consists of a rigid bar supported at a point known as the fulcrum. A force called the effort force is applied at one point on the lever in order to move an object, known as the resistance force, located at some other point on the lever. A common example of the lever is the crow bar used to move a heavy object such as a rock. To use the crow bar, one end is placed under the rock; the bar is supported at some point (the fulcrum) close to the rock. A person then applies a force at the opposite end of the crow bar to lift the rock.

In the MOVEMENT chapter, it explains that torque can be increased by applying a force as far away from the fulcrum, or axis of rotation, as pos-

What Earth-moving statement did Archimedes make about the lever?

Sometime in the third century B.C., while conducting research on levers, Archimedes stated, "Give me a firm spot on which to stand, and I will move the Earth." Archimedes was referring to the use of levers; theoretically, with a long enough lever and a non-Earthbound place to rest a fulcrum, one could move the Earth.

sible. Therefore, the longer the lever-arm a person uses, the less force she needs to apply to that lever in order to lift up the object. By using long enough levers, extremely heavy objects can be moved.

What is a **class-one lever**?

Class-one levers are levers in which the fulcrum is somewhere between the point of applied force and the point of exerted force. Class-one levers consist of an effort arm where force is applied by the user, a pivot point called the fulcrum, and a resistance arm where the object to be lifted or moved is placed. For example, a child's teeter-totter is a class-one lever. When a child is lifted in the air, he is sitting on the resistance arm; the child with his feet on the ground is sitting on the effort arm. Other class-one levers are rowing oars and crow bars.

Are **scissors and pliers** levers?

A pair of scissors is actually two first-class levers working simultaneously on a common fulcrum. For scissors, each shear is a lever, and when force is applied to the handles, the two levers rotate over the shared fulcrum, passing very close to each other and enabling their sharp edges to cut material.

What is a **second-class lever**?

A second-class lever is a lever in which both the effort and resistance forces are on the same side of the fulcrum, with the resistance force between the fulcrum and effort force. An example of a second-class lever is the wheelbarrow, invented by the Chinese in A.D. 400. Here the fulcrum is the wheel, the resistance is the dirt, and the effort is the handle.

What are some examples of **third-class levers**?

Like a class-two lever, a class-three lever has both forces on the same side of the fulcrum, but in this class, the resistance force is farther away from the fulcrum than the effort force. This causes a mechanical advantage of less than 1. Although more force is needed, it does increase the speed of the lever. Instead of making it easier to lift an object, it allows your hand to move less than it would using a first-class lever.

Examples include fishing poles, where the resistance is the fish, and the effort force—the hand—is closer to the fulcrum. The fulcrum in this case would be where the pole is resting against the body of the person fishing. Another example of the third-class lever is a baseball bat. The hands are extremely close to the fulcrum, which is at the base of the bat, but the resistance force, where the ball hits the bat, moves quickly in order to hit the ball.

PULLEYS

What is a **pulley**?

To raise objects to a level where an inclined plane would be impractical, a series of pulleys and ropes can be used to gain mechanical advantage. A pulley is a wheel (with an axle or bearing in its center) on the rim of which rides a rope or cord. The ancient Assyrians used a single pulley to lift objects up to roof tops. Using one fixed pulley provides no mechanical advantage, but does allow one to pull from underneath instead of pushing, or pulling from above. Single fixed pulleys simply change the direction of a force.

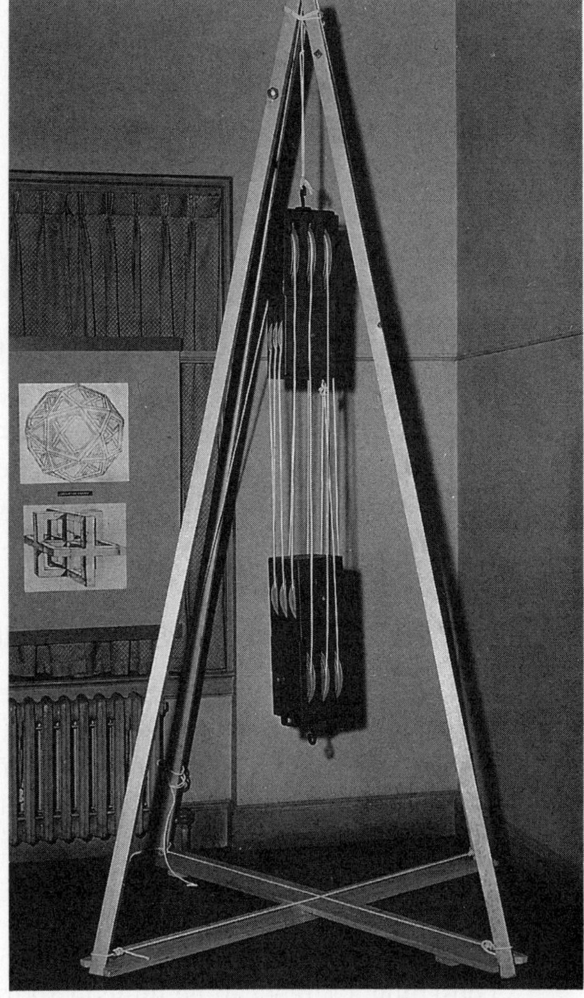

A block and tackle circular pulley.

What did the Greeks and Romans do to **improve the mechanical advantage** of the pulley?

The Greeks and Romans used multiple pulleys, around which were looped a single rope, to raise objects with less effort. Generally, the greater the number of pulleys, the higher the mechanical advantage. In one of the earlier systems, the Romans used five pulleys and achieved a mechanical advantage of five to one. This system of multiple pulleys is called a block and tackle.

To better accommodate the block and tackle system, ancient engineers constructed mechanical cranes that attached to the top pulleys to lift objects higher than building roof tops. Archimedes even designed a sailing ship equipped with the block and tackle system of multiple pulleys so the captain could sail a ship solo.

Wheels and Axles

What is a **wheel and axle**?

A wheel is a circular disk or rim attached to a central rod, called an axle, about which the wheel can turn. The steering wheel in a car is a wheel and axle. The section that we place our hands on and apply torque is called the wheel, which turns the smaller axle. The larger the diameter of the wheel, compared to the diameter of the axle, the greater the mechanical advantage.

What culture **developed** the wheel and axle?

Historians believe that in the first or second century B.C., Europeans may have been the first to develop a device called the rotary quern. This device consisted of a crank attached to an axle, which was used to turn a circular millstone used to grind corn. This was the first sign of a wheel and axle machine.

How is a **screwdriver** a wheel and axle?

Loosening a screw can be difficult to impossible to do with your bare hands. A screwdriver aids this process by applying more torque to the turn. The handle, preferably thick, is the wheel, while the metal shaft is the axle. The larger the diameter of the handle, the less the force that is needed to tighten or loosen a screw.

Gears

What were **treadmills** and how did they help develop gears?

Treadmills were used at milling stations around Europe shortly after the wheel and axle was developed. It required a man or a few men to walk

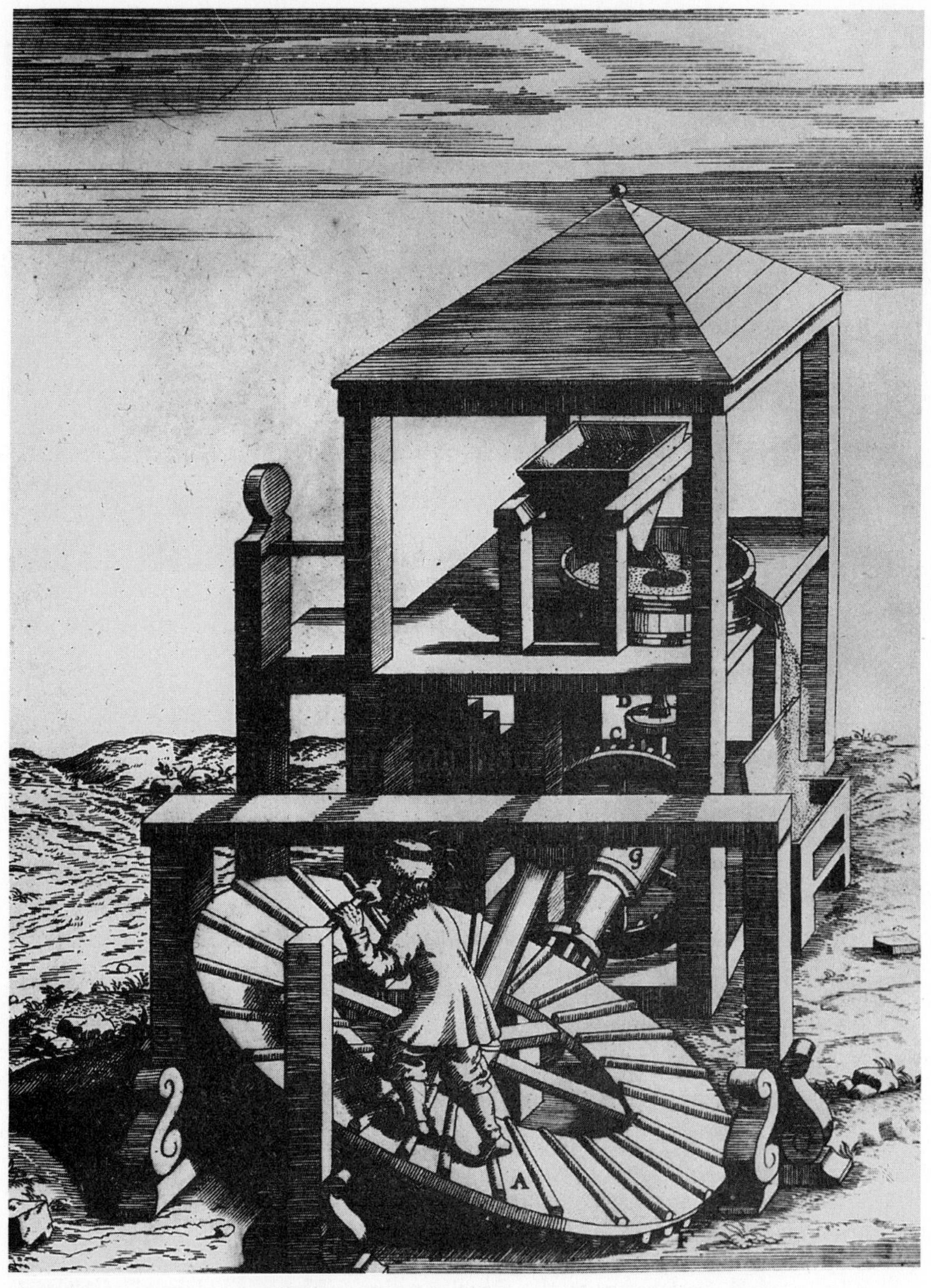

A horizontal treadmill used to turn gears.

Gears, including a worm gear on the left center of the photo.

inside what we call today a squirrel-wheel. Squirrel-wheels are round wheels that pet rodents use for exercise. The squirrel-wheel was placed vertically so the men could remain upright while walking. The axle attached to the wheel was horizontal, but to grind corn, the axle needed to be vertical to turn the mill. As a result, gears were invented. Gears were not originally used for mechanical advantage, as the wheel and axle was, but for directional changes, which allowed treadmills to be used for a more efficient milling of corn.

A variation of the squirrel-wheel was a horizontal platform that sloped from its center, with ridges for treads (see illustration).

Archimedes studied the lever, but who studied **gears**?

In the A.D. first century, the Greek engineer Hero (or Heron) of Alexandria, Egypt, documented and described all the known gears at the time in a book called *Mechanica*. Hero also developed the first, quite primitive steam engine, and had several breakthroughs in the field of geometry.

What are **gears**?

Gears, a by-product of the wheel and axle, can produce huge mechanical advantages in machines. Gears can change orientations for axles as the first gears did in milling stations, or they can multiply forces, and even help run precise time-keeping devices.

Gears consist of circular wheels with teeth carved into them. The mechanical advantage of a gear system is determined by the number of teeth on the driven gear divided by the number of teeth on the gear driving it. To increase mechanical advantage, the driver gear is smaller and has fewer teeth than the gear being driven. This is called gearing down. Gearing up results in a mechanical advantage of less than 1, but increases the speed of the gear train. When gearing up, the driver gear must be larger and have more teeth than the driven gear. The resulting gear train is not stronger, it is faster.

What is the **Antikythera**?

The Antikythera is a mechanical calendar, consisting of twenty-five interconnected bronze gears. It is estimated that the Antikythera, which was recovered from an old shipwreck, was made in Rhodes during the first century B.C. The significance of the device is that it demonstrates the high-level mathematical and engineering technology present two thousand years ago.

ENERGY

When a **falling object stops,** what happens to its energy?

Because the law of the conservation of energy states that the energy present in a system is always constant, an object that falls and hits the ground should bounce back up to where it started. Although it would seem that the kinetic energy should be transferred back into potential energy, some of the energy, upon hitting the ground, is transferred into heat energy. In fact, some kinetic energy is also transferred into heat from the friction between air molecules and the object. Although energy is conserved, it is not necessarily transferred between mechanical energies; instead, a great deal of energy is changed into heat. This is known as "inelastic collision."

Do simple machines lose energy to **friction**?

An ideal machine would get just as much work out as is put in; however, such machines don't exist. A great deal of energy is transferred into heat energy from frictional forces taking place between moving parts. Most machines are measured for efficiency, which indicates how much work was put into a system compared to the amount of work performed by the machine.

Energy Efficiency

How energy efficient is the average **automobile**?

Out of the total work or energy put into a car through the chemical potential energy of the gasoline, only about a quarter of the energy goes into moving the vehicle. The other 75% of the energy is transferred into other energies that are not useful for moving the automobile. For example, the friction of moving parts changes kinetic energy into thermal energy, which only increases the temperature of the engine. The rest of the thermal energy leaves the vehicle as hot exhaust through the tailpipe. Engineers are constantly working on methods of increasing the efficiency of automotive engines.

Have **emission levels** from automobiles decreased over the past few years?

Carbon monoxide emissions account for nearly 60% of airborne pollutants—and of that 60%, 80% is from automobile emissions. However, the 60% figure has been on a slow decline over the past fifteen years or so, due principally to emission controls, energy conservation, and alternative energy programs.

Has the energy efficiency of automobiles **improved** since the 1970s?

Americans are buying more cars per household and driving more miles; however, the average fuel economy of automobiles has increased, resulting in less gas used per mile for each vehicle and fewer polluting emissions. The following is a comparison outlining the average number of miles traveled per each car in the U.S., fuel consumption, and fuel economy, over the past twenty-five years.

Year	Distance (miles)	Fuel consumption (gallons)	Fuel economy (miles per gallon)
1970	10,271	760	13.5
1975	9,690	718	13.5
1980	9,141	590	15.5
1985	9,560	525	18.2
1990	10,548	502	21.0
1995 (estimates)	11,000	495	22.2

Energy Production and Consumption

What countries are the **largest producers of energy** in the world?

The largest producers of energy in the world are:

Rank	Country
1	United States
2	Russia
3	China
4	Saudi Arabia
5	Canada
6	United Kingdom
7	Iran
8	Norway
9	India
10	Venezuela

Together, Russia and the United States produce nearly one-third of the energy in the world.

What countries are the **largest consumers of energy**?

The United States consumes over 28% more energy than it produces. That accounts for nearly a quarter of the energy consumed throughout the entire world, while the population of the United States accounts for only 5% of the world's population.

Rank	Country
1	United States
2	China
3	Russia
4	Japan
5	Germany
6	Canada
7	India
8	United Kingdom
9	France
10	Italy

What are the **average yearly costs** of some general home **appliances**?

Home climate control and appliances account for approximately one third of the energy consumption in the United States. The average cost for energy is approximately $.12 per KwH (kilowatt-hour), but varies throughout the country. The following is a listing of home appliances, their typical usage, and the cost for one full year.

Appliance	Energy (KwH)	Annual Cost @ $0.12 / KwH
Television (8 hours per day)	1000	$120
Stove with Oven	1000	$120
Washer	150	$18
Dryer	1000	$120
Refrigerator	1200	$152
Frost-free Refrigerator	2000	$240
Hot-water Heater	5000	$600
Window Air-conditioner (if used year 'round)	1500	$180

Which states consume the most energy in the United States?

Overall, the largest consumer of energy in the United States is Texas, followed by California, Ohio, and New York. However, the largest consumer per capita in the U.S. is quite different. The states that consume the most energy per person are Alaska, Louisiana, Wyoming, Texas, and North Dakota. It's not surprising that the people of Alaska would have the highest per capita usage, given their greater need for heating.

The lowest states for overall consumption are Rhode Island, South Dakota, and Vermont. Per capita, however, the lowest consuming states were California, Hawaii, and New York.

ALTERNATIVE ENERGY

What does **alternative energy** mean?

Alternative energy is any form of energy that doesn't originate from fossil fuels. Forms of alternative energy include nuclear energy and renewable resources such as hydroelectric, geothermal, biomass, solar, and wind energy. Currently about 80% of our energy consumption comes from fossil fuels such as coal, natural gas, and petroleum.

What are **hydroelectric, geothermal, biomass,** and **wind energy**?

Hydroelectric energy is generated by sending high-speed, high-kinetic energy water through a series of turbines that changes the water's kinetic energy into usable electrical energy. Geothermal energy takes energy from within the Earth, typically in the form of steam, and transfers the high-energy steam into usable electrical energy. Biofuels or biomass includes the burning of wood and wood products along with waste and

Power-generating windmills in Altamont, California.

landfill gases to produce electrical energy. Finally, whereas hydroelectric uses the kinetic energy of the water to generate electricity, wind energy uses numerous spinning windmills to change the kinetic energy of the air into usable electric energy.

How much energy does the Earth receive from the **sun,** compared to what the Earth has in **fossil fuels**?

The solar energy received from the sun is more than 15,000 times what the Earth consumes each year. It is estimated that each year the Earth receives ten times more energy from the sun than the entire Earth's supply of fossil fuel.

Who invented the first **solar-powered collector and motor**?

It was a French mathematician by the name of Augustin-Bernard Mouchot who first successfully attempted to collect the sun's energy. Unfortunately, coal was so abundant and easy to use that there was little interest

A solar-energy system at the National Renewable Energy Laboratory, in Golden, Colorado.

in Mouchot's invention. It wasn't until the energy crisis of the 1970s that people throughout the United States and the world began to realize that fossil fuels were not limitless. It took over one hundred years since Mouchot's experiments for the United States to look seriously into the use of solar energy. Although there is a great deal of enthusiasm for solar power, today it accounts for only 0.1% of the energy production in the U.S.

How much energy is currently produced from the various **energy sources**?

Energy source	Percentage (%)
Wind farms	0.1
Solar	0.1
Geothermal	0.5
Biofuels (wood, landfill gases, agricultural waste)	4.3
Hydroelectric	5.0
Nuclear	9.9
Fossil Fuels (natural gas, coal, crude oil)	80.0

CENTER OF MASS

When tossed in the air, why do **hammers** appear to **wobble end over end**?

If a baseball is tossed into the air, the ball follows a smooth parabolic (curved or arch-like) path as defined by the laws of physics. However, if the same is done to a hammer or wrench, it appears to wobble throughout its motion. The wobbling is caused by the hammer's non-uniform distribution of mass.

The center of mass in any object is defined as the average location of its mass. Because the mass is distributed evenly throughout a baseball, the center of mass is located in the center of the ball. However, for an object such as a hammer, the center of mass is not directly in the middle. Since more mass is located in the metal head of the hammer, the center of mass is closer to that point.

The laws of physics state that the center of mass follows a parabolic curve when tossed in the air. Indeed, although the ball and the hammer do not appear to have similar motions, their centers of mass do. If you watch closely both the center of a baseball and the center of mass of a hammer, you will see that they both follow parabolic paths when thrown.

Despite the apparent wobbling of a wrench when spun on a horizontal surface, the wrench's center of mass moves in a straight line.

What is the difference between **center of mass** and **center of gravity**?

Whereas the center of mass is the average location of an object's *mass,* the center of gravity is the average location of an object's *weight,* where the weight is the mass times the acceleration due to the gravitational field. For most objects, there is no difference between the center of mass and the center of gravity. However, for larger objects such as planets, the center of mass may be slightly different from the center of gravity. For example, the moon is uniform and its center of mass is in its center. However, since the Earth exerts a gravitational force on the moon, and because gravity is dependent on mass and distance, the far side of the moon will not experience the same gravitational force toward the Earth that the near side does. Therefore, since the near side of the moon experiences a greater force than the far side, the center of gravity for the moon is just a bit closer to the Earth than the center of mass.

For objects on Earth, the difference in gravitational pull from one side of an object to the other is virtually nothing. Therefore, for normal everyday use, the center of mass and the center of gravity can be used interchangeably.

Where is the center of mass for a hula-hoop?

Just as the mass in a baseball is evenly distributed throughout the ball, the same holds true for a hula-hoop. However, there is one important exception. The hula-hoop is a loop, and therefore has no mass located in its center. However, the center of the hoop is still its average location of mass, and therefore, its center of mass.

What is a **support area**?

A support area is responsible for helping to keep objects upright. If you were to imagine a line drawn on the ground outlining your feet and the area between them while standing, the area inside the line would be called your support area. For example, the support area of a building is its foundation on the ground. The support area of a bicycle would be the area between and including where the tires touch the ground.

Why is a support area critical to remaining **upright**?

If an object is unstable, it means that it could easily tip over. In order to tip over, the object's center of mass must move horizontally beyond its support area. If this happens, there is nothing to support the object anymore, and it tips.

In order to make something more stable, a larger support area is needed, which makes it more difficult for a force to push the center of mass beyond its support area. For example, the Eiffel Tower in Paris was designed with a wide base and a thin top. The wide base gives it a huge support area and makes it extremely difficult to knock the Tower over. If it did not have such a wide support area, the Tower could topple if the wind blew hard enough for the center of mass to move beyond its foun-

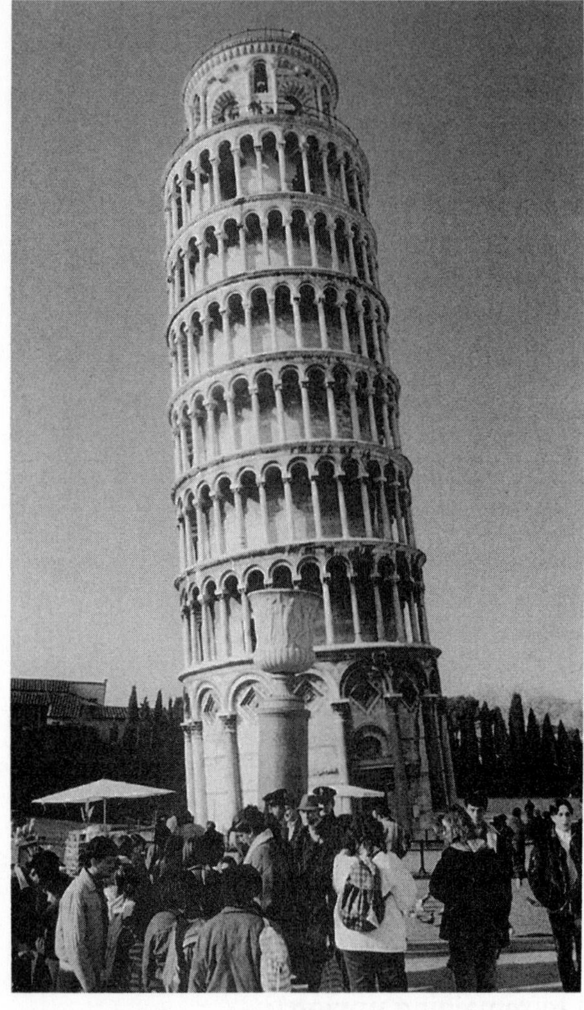

The Leaning Tower of Pisa.

dation. This is the problem found in many radio towers; although they do not have large support areas, they do have strong cables—called guy wires—that help support the structure.

Why are **football players and wrestlers** taught to **get down low** when blocking or making a move on someone?

When confronting another athlete in a physical manner, the best thing to do is to get down low and spread your feet apart. This enlarges your

A monkey's tail helps to keep the monkey's center of mass over its feet (see next page).

support area, thereby making your body more stable and difficult to knock over. Your opponent, in order to knock you over, would have to exert a force to lift up your center of mass and then push you over. If you simply stood upright, with your feet close together, your support area would be extremely small and your opponent would have less difficulty pushing you and your center of mass beyond your support area.

Why hasn't the **Leaning Tower of Pisa** fallen over yet?

The 56 meter (185 foot) Leaning Tower of Pisa has been tipping over since construction began in A.D. 1173. It is leaning because its foundation is not firmly fixed to the Earth; although the foundation is over 3 meters deep, it was not built into solid bedrock, which would have prevented the tilting. Despite its 5.18-meter tilt, the tower still stands because its center of mass is still above its support area or foundation. The tower, now closed to the public, continues to tilt 1.25 millimeters per year, and will continue to do so until the center of mass is no longer above its supporting foundation or, more likely, until the tower walls or foundation structurally fail. The tower is in a shear stress due to the leaning, and the walls and foundation are designed to distribute stress straight down, not sideways too. Most likely the walls will fail one day and the tower will "buckle" in the middle somewhere.

Why do **monkeys** have **tails**?

Biologists may see many uses for or reasons why monkeys have tails, but physicists see the tail of a monkey as a wonderful balancing tool. The tail is used primarily to help keep the monkey's center of mass over its feet. For example, if a monkey stands on a tree branch and reaches out to grab a banana, its center of mass would probably move past its feet, causing the monkey to fall. In order to combat this problem, a monkey will extend its tail out behind its back, keeping the average location of its mass over its feet. Birds, squirrels, and many other animals with tails do the same thing.

STATICS

What does it mean to say that an object is **static**?

Static means "not moving." In static electricity, charges do not flow, but remain in one location until a force comes along to move them. In the fields of engineering and mechanical physics, to be static means that the object does not move. When static, all the forces acting on a body must cancel each other out so the object does not move and the "net" or total force on the body is zero.

Why are we static when **sitting in a chair**?

As long as you are sitting in a chair and not moving (relative to the Earth), you are static. That means that the chair is supporting you with an upward force that is equal to your weight. You would remain static until some external force came along to change your motion.

What is the name of the **supporting force** from the chair?

Another term for a supporting force is "normal force." The normal force is always directed perpendicularly out of the surface. The normal force of a chair is straight up if the chair is on a level surface, while the normal force of an incline would be perpendicular to the surface of the incline, and not perfectly vertical. The term "normal" was derived from the geometrical name for a 90° angle.

What is a **tension force**?

A tension force is a force that attempts to pull things apart; it is often found in taut ropes, cables, and wires. When a child hangs from a chin-up bar, he is experiencing a tension force in his arms. When hanging from two arms, the tension in each arm is half his weight. If only one arm is used, it is more difficult to hold on because the single arm now experiences a tension force equal to the child's entire weight.

Why is it so difficult for **gymnasts** to perform the **iron cross** on the rings?

If you have ever tried the iron cross on the rings, you would know that unless you are extremely strong, there is no way to achieve such a static position. The reason it is so difficult is the huge force that is needed to suspend one's body from the rings. Physics says that the closer a tension wire, rope, or in this case, arms, get to a horizontal position, the more force is needed to keep the object suspended. In other words, the easiest way to hang something is from vertical wires or arms; the closer the wire gets to horizontal, the more tension there is.

Take, for example, a clothesline. The clothesline is horizontal, which means that it experiences huge tension forces even if something relatively light is draped over it. A clothesline could be replaced with vertical strings and still support the same clothing with a lot less tension. But one problem would remain—to what would you tie the vertical strings?

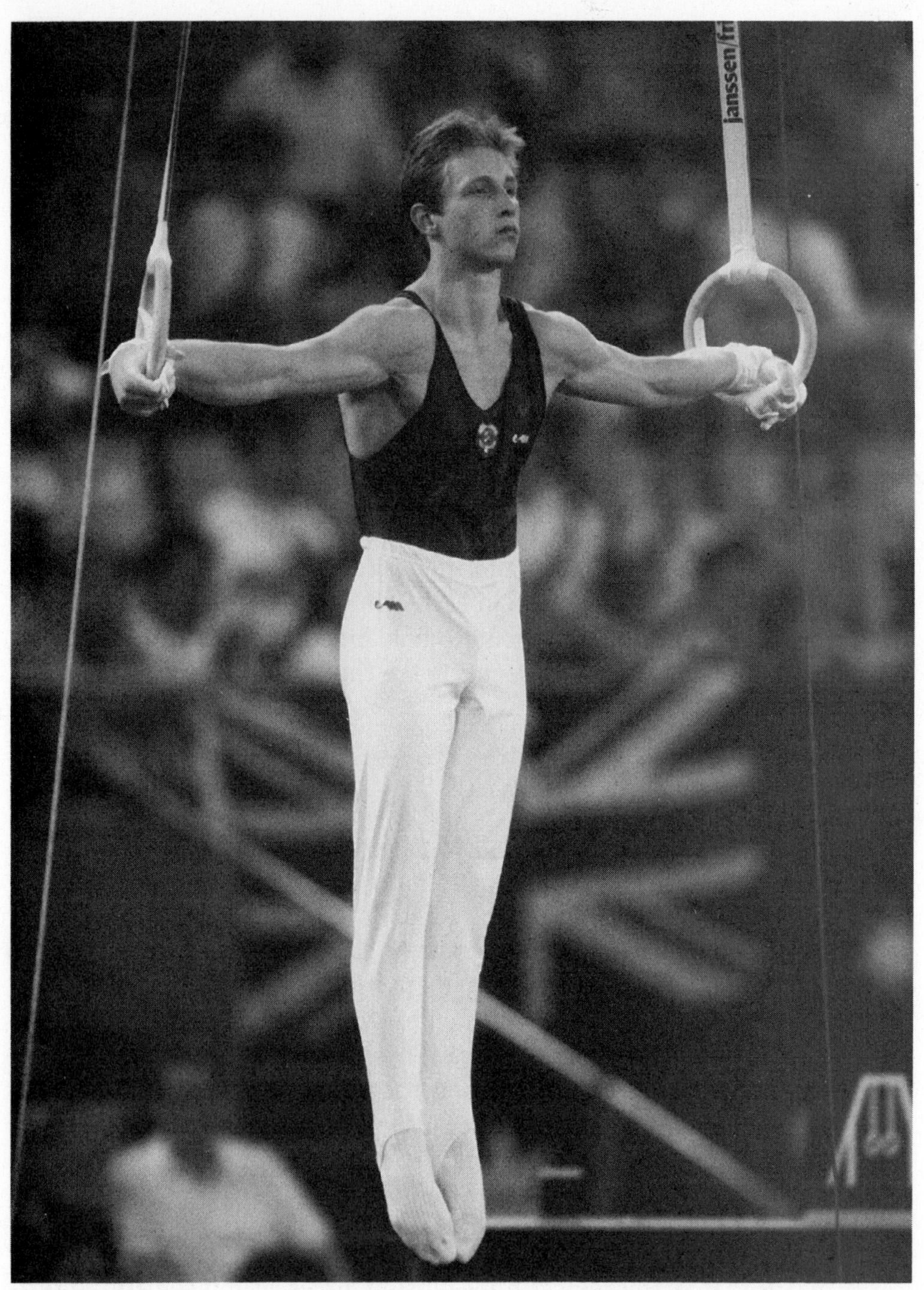

Gymnast performing the iron cross (see previous page).

What is the **opposite** of tension force?

Tension is a pulling force, and compression is a pushing force. If you are sitting down right now, you are exerting a compression force on your backside and on the chair. Even while standing, you are compressing your legs and the ground beneath your legs. Wires and cables do not deal well with compression forces, so materials such as steel, iron, and concrete are used for building foundations and support beams that are needed to withstand large compressions.

What is **shear**?

One of the best examples of a shearing force can be found in scissors. Scissors, also known as shears, use each of their blades to move the object it is attempting to cut in opposite directions. A shearing force does the same thing. Earthquakes often cause structures to experience significant shearing forces. Pictures of torn-up roads after earthquakes show how one side of a street moved one way while the other side of the street moved in the opposite direction, tearing each other up as they moved past one another.

What **other major force** can be experienced by structures?

A torsion force is responsible for twisting structures. Bridges and towers use cross-supports to prevent such forces from damaging the structures.

BRIDGES AND OTHER "STATIC" STRUCTURES

What are some of the **major classifications** of bridges?

There are four fundamental styles of bridges, and today's civil engineers use many variations of those styles. These four styles are the beam, cantilever, suspension, and arch bridges. There are hardly ever two identical

What is the difference between a dead load and a live load?

In order remain static, bridges (and all structures, for that matter) must be able to withstand loads placed on them. A load is simply an engineering term for force. Dead load is the weight of the bridge or structure itself. The live load, however, is the weight and forces applied to the bridge as a result of the vehicles and people that move across the bridge at any one time. Of course, in order to be safe, engineers account for much higher live loads than would normally occur.

bridges in the world, because all bridges must be engineered according to such specifics as geographical and geological conditions, cost, aesthetics, as well as how often and how much the bridge will be used.

How can **suspension bridges** be so long?

The twenty longest bridges in the world are all suspension bridges. Suspension bridges are able to span huge distances because the long cables suspending or holding up the roadbed are draped over a set of tall vertical towers. These towers prevent the mid-section of the cables and roadbed from plummeting into the water. However, the ends of the long cables must be anchored into the ground at each end of the bridge, otherwise the weight of the roadbed and mid-section of cabling would be enough to make the towers cave in toward the center of the bridge. The graceful lines of the suspension have wonderful aesthetics, as well.

What is the **longest bridge** in the world?

As of this writing, the longest suspension bridge in the world is nearing completion in Kobe, Japan. Its name is the Akashi Kaikyo, and it will

Verrazano Narrows suspension bridge linking Staten Island and Brooklyn.

span a distance of 1,990 meters (6,527 feet). This $3.3 billion bridge is impressive not only in its size but because it has already weathered the 7.2 Richter earthquake that killed 5,000 citizens of Kobe in January 1995. The only damage sustained by this incredibly engineered bridge was that one of the piers and anchorages shifted a little less than one meter. This high-tech bridge—whose towers are about the same height as the Eiffel Tower in France—uses springs and pendulums within the massive vertical towers to counteract dangerous bridge movement produced by seismic activity. These high-tech mechanisms move against the motion of the bridge, stabilizing it and keeping drivers on the bridge relatively safe.

Where is the **longest bridge** in the **United States**?

The longest bridge in the U.S. is also a suspension bridge, and ranks as the sixth longest bridge in the world. It is the Verrazano Narrows Bridge, between Staten Island and Brooklyn, New York. This bridge, completed in 1964, spans 1,298 meters (4,260 feet).

The Pont du Gard arch bridge in Nimes, France.

How does an **arch bridge** support the weight on top of it?

Arch bridges are known for their stability. The force applied to an arch is translated into a compression force that is carried outward from the center crown of the arch to its bearings, or abutments, on either side of the arch. The Romans were famous for the arches used in their massive aqueducts. One such aqueduct, the Pont du Gard, was completed in 18 B.C. and was used to carry water a length of 270 meters (886 feet) over the Gard River in southern France.

What is a **cantilevered bridge**?

Typical cantilevers consist of an array of steel beams that shoot out from a base to support a roadway. The roadway depends completely on the strength of the cantilever. Most cantilevered bridges have two or three cantilevered sections. These bridges are no longer constructed due to their high cost and complexity. Examples of cantilevered bridges in the United States are the Commodore Barry Bridge in Chester, Pennsylvania, the Greater New Orleans I and II in New Orleans, Louisiana, and the Gramercy Bridge in Gramercy, Louisiana.

The Firth of Forth cantilever bridge in Edinburgh, Scotland.

What was the **first type of bridge**?

The first type of bridge ever used was a beam bridge. This bridge was probably just a fallen tree that was used to cross a ravine or a small stream; the tree was probably supported by the river bed or by a group of rocks. Beam bridges consist of a horizontal roadbed supported by vertical piers that are planted in the ground. Although beam bridges can be extremely strong, they are not effective over long distances.

What is the **newest hybrid** of bridges?

One of the newest, prettiest, and most economical bridges is the cable-stayed suspension bridge. With its sleek lines and thin roadways, it is the perfect bridge for most mid-span designs. The Tatara Bridge, in Onomichi, Japan, will be the longest cable-stayed bridge in the world, with a span of 890 meters (2,919 feet). Cable-stayed bridges suspend the roadbed by attaching multiple cables directly to the deck supporting the roadbed. These cables are then passed through a set of tall vertical towers and attached to abutments on the ground. Such engineering meth-

The Dame Point cable-stayed bridge in Jacksonville, Florida (see previous page).

ods reduce the need for heavy, expensive steel and the massive anchorages that are needed to support suspension bridges.

How tall is the **Eiffel Tower**?

The Eiffel Tower, built for the 1889 Paris Exhibition, stands 321 meters tall (1,052 feet). It's named for the tower's designer, Gustave Eiffel, who wanted to design a contemporary structure for the 100th anniversary of the French Revolution.

What is the **tallest building** in the world?

The tallest building, until recently, was the Sears Tower in Chicago, Illinois, built in 1974. It stands a full 443 meters (1,453 feet) high. However, three new skyscrapers in Asia will surpass the height of the Sears Tower. The Petronas Tower in Malaysia is 452 meters (1,483 feet) high, and was finished in 1996. The Chongqing Tower in China and the Shanghai World Finance Centre in China will be completed around the

The Shanghai World Finance Centre will be 460 meters (1,509 feet) tall.

turn of the century. The tallest of them all, the Shanghai World Finance Centre, will stand 460 meters (1,509 feet) high—almost half a kilometer into the sky.

What is the **tallest structure ever built**?

The Warszawa Radio Tower on the outskirts of Warsaw, Poland, was the tallest structure ever built. Although it needed to be supported by long cables to keep it up, the tower reached 646 meters (2,119 feet) into the

What makes a building a skyscraper?

In order to be a skyscraper, a building must be supported by an internal iron or steel skeleton, instead of being held up solely by load-bearing outer walls. Skyscrapers have been useful and economical in crowded cities because they take advantage of the more abundant vertical space that is available; also, skyscrapers use iron and steel, which are easier to work with, weigh less, and are more compact than concrete.

sky. Unfortunately, in August of 1991, the tower came crashing down during repair work.

The tallest structure ever built that is still in existence is the KTHI-TV tower in North Dakota, which stands—with the help of cables—629 meters (2,063 feet) tall. The tallest self-supporting tower in the world is the Canadian CN Tower, whose tip is located 553 meters (1,815 feet) above the city of Toronto.

What is a **fluid**?

Any liquid or gaseous material that can flow is considered to be a fluid. Fluids play an important role in every aspect of our lives, including breathing, flying, and swimming. There are two main areas of fluid study. The field that studies fluids in a state of rest is called static fluids, and the field that analyzes the movement of fluids is called fluid dynamics.

STATIC FLUIDS

WATER PRESSURE

What does it mean to say that **water seeks its own level**?

The surface of water placed in a single container (a glass or a bathtub or a lake) will remain at the same level relative to the Earth on both sides of the container. Adding water to one side will only make the entire level uniformly rise; there can never be one section of the glass or tub or lake that is at a higher elevation than another section. Water and all liquids rest at the same level.

A spherical (and cheerful) water tower in West Branch, Michigan.

Why are **water towers** needed on tall buildings?

Water flows from areas of high elevations to low elevations. This can cause problems in tall buildings and very flat areas of the country. In order to have enough water pressure to reach the top of tall buildings, water towers are often placed on the roofs. These water towers (which are initially filled by using pumps) are not used to store water, but instead to provide enough pressure to lift water to upper floors in buildings. Since water seeks its own level, if a water tower is on a roof, the rest of the building's water will push upwards to rest at the same level. This "push" is the water pressure for the building.

Why are many municipal water towers **spherical in shape** and placed on high towers?

The reason why a spherical holding tank is placed on top of a water tower is to maintain enough water pressure for an entire town or community. The depth of water determines the liquid pressure. Since a hold-

What are the bends?

When diving in deep waters, the pressure from the water above is much greater than it would be near the surface. If a diver swims to the surface too quickly, the rapid change in pressure can cause nitrogen to show up in the blood.

Nitrogen, under normal atmospheric pressure, is nearly insoluble in blood. Under pressure, the solubility increases. Thus, as a diver goes deeper the blood holds more and more nitrogen, which dissolves in the blood during gas exchange in the lungs. As the diver ascends, the pressure decreases, and hence the blood is now supersaturated with nitrogen. The supersaturated nitrogen forms bubbles as it comes out of solution in the blood, or cells. These bubbles collect in joints, arteries, and other places. They cause pain, and can rupture cell walls and block the flow of blood—hence oxygen—to cells, causing injury or even death.

The best way to avoid the bends is to rise to the water surface slowly, allowing the liquid pressure from the water to gradually decrease and hopefully prevent any physical damage.

ing tank is up high, it has a great deal of potential energy and places a lot of pressure on the water in the rest of the water network.

The need for great height of the tower may make sense, but why have a spherical holding tank on the top, instead of constructing a tower that looks more like a traditional silo?

Pressure depends solely on the depth of the water. If the same amount of water were in a silo-shaped holding tank, the depth would not be as great, reducing the water pressure. However, when water is held in a spherical holding tank at the top of the tower, the amount of water needed is reduced, yet there is still enough water to create a large potential energy to maintain pressure. During periods of droughts or high consumption, the water level in the spherical tower may drop, but water pressure will remain relatively high.

Where is the **water pressure greater,** in a lake 20 meters deep or in the ocean at a depth of 10 meters?

Pressure is defined throughout physics as the force divided by its surface area. Although the ocean contains a lot more water than a lake, it is the depth or weight of water directly over a diver that defines the amount of pressure the diver experiences. Therefore, a diver who is 20 meters below the surface of a lake will experience more pressure, in fact, twice the pressure, than the ocean diver experiences at 10 meters.

Why do your **ears hurt** when you dive to the bottom of a swimming pool?

Just as the weight of the air above us creates atmospheric pressure, the weight of water creates liquid pressure. Close to the surface of a pool there is very little water that can push down and compress the water. However, the further a person dives below the surface, the greater the water pressure from above. If a diver is close to the bottom of a pool, his body can actually feel the increased pressure. The eardrums are especially sensitive to the increased pressure, for they do not have the reinforcement that the diver's skin has. In fact, your eardrums can usually feel pressure when diving just 5–10 feet below the surface of the water.

Why are **dams** thicker at the bottom than at the top?

Dams hold back bodies of water, and water pressure increases with the depth of the body of water, so the pressure from the water wanting to escape is greater at the bottom of the dam than at the top. If holes were bored near the bottom, middle, and top of a dam, the longest horizontal stream of water would fire out through the bottom hole because the water pressure is greater there.

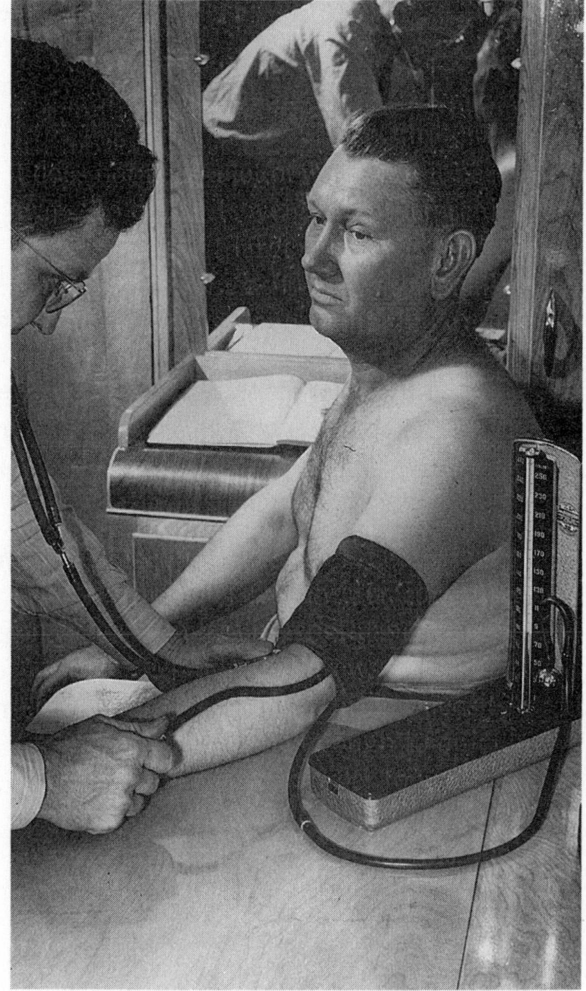

Man having his blood pressure measured with a sphygmomanometer.

Blood Pressure

What does it mean to measure your **blood pressure**?

Blood pressure is the pressure your blood exerts on the walls of your arteries. The fluid dynamics of blood play a major role in blood pressure. The heart is the pump that moves the blood throughout the body, with vessels carrying the blood to different sections of the body. These vessels can feel a great deal of pressure, because the movement of blood from larger arteries to smaller arterioles and capillaries can create back-ups, which increases the pressure against the walls of the vessels.

The device used to measure blood pressure is the sphygmomanometer. It is placed around the upper arm, inflated, and then deflated, while a meter measures the pressure passing through that section of the arm.

Why is your blood pressure taken from your **upper arm**?

Liquid pressure is dependent on the depth of the fluid. Since blood pressure can't be measured around the heart, and the depth of the fluid must be the same as the heart, doctors and nurses need to find a location at the same depth as the heart. A convenient location at that level is your upper arm. However, when lying down, your blood pressure can be taken just about anywhere, since most of the blood is at the same vertical level as the heart.

The amount of blood pressure on the outer walls of your arteries and veins can actually be seen and felt when you are standing on your head. Since there is a lot of blood above the head when upside down, the vessels experience a great deal of pressure, something that your feet may be used to, but your head is definitely not. In fact, the pressure on the walls of the blood vessels can be so great that the veins in a person's head and neck can be seen protruding from the skin. This is known as a head rush.

Atmospheric Pressure

How is **atmospheric pressure** similar to **liquid pressure**?

Atmospheric pressure, which is the pressure from a gas, acts the same way liquid pressure does. The only difference between atmospheric and liquid pressure is that gases are less dense than liquids and therefore apply less pressure on a person or object. For example, the Earth's atmosphere extends approximately 45 to 50 kilometers above the ground. The amount of air pressure on an area of one square meter is about 100,000 newtons per square meter. If, however, we were talking about water instead of air, the force on one square meter would be much greater.

What is a barometer?

A barometer is a device used to measure air pressure. There are two major types of barometers, the mercury barometer and the aneroid barometer. Galileo's secretary, Evangelista Torricelli, developed the mercury barometer in 1643. It consists of a thin glass tube about 80 millimeters long, which is closed at the top, filled with liquid mercury and placed upside down in another mercury-filled dish. Depending upon the atmospheric pressure pushing on the mercury in the dish, the level of mercury in the tube will rise or fall because there is no air in it. By measuring the height of the mercury, which would usually be between 737 and 775 millimeters high, the relative pressure of the atmosphere can be measured.

If the **air pressure** is 100,000 newtons per square meter, why don't we get **crushed**?

According to Newton's Third Law, we exert an equal pressure on the atmosphere. Since our bodies also have air inside them, the air inside our bodies is at the same atmospheric pressure as the air outside our bodies. Therefore, the pressures are equal and we can move quite freely in our atmospheric environment.

The same cannot be said for water and divers. Divers far beneath the surface can feel a 100,000 newtons per square meter pressure because the air inside the diver's body is not at the same pressure as the water outside the diver. The only way to appease the pressure of the water would be for the diver to swallow the highly pressurized water. Not a good idea.

There is a second method of equalizing pressure without swallowing water. At extreme depths, air can be pressurized to the same pressure as the water, if given enough time. Divers in such high-pressure chambers would be able to enter the water at such depths without being crushed. A fictional example of such an occurrence is found in the movie *The*

Abyss. The diving module had an opening to the sea, so the air pressure had to be equal to the water pressure, or the water would have flooded the module. The divers could then jump directly into the water and swim around.

What is an **aneroid barometer**?

The more common barometer is the aneroid barometer, in which atmospheric pressure bends the elastic top of an extremely low-pressure drum; by measuring the amount the top bends, a measurement of atmospheric pressure can be determined. The aneroid barometer is often used in airplane altimeters to measure altitude. Since atmospheric pressure decreases as altitude increases, the aneroid barometer is an ideal instrument to use. It is much safer than the mercury barometer, because mercury is poisonous; a mercury barometer requires the mercury in the dish to be unenclosed, making it vulnerable to spilling.

What happens to a **balloon** when it is submerged in **water**?

When an air-filled balloon is placed underwater, the water (which has higher pressure than the air) exerts a force on all sides of the balloon. The pressure from the water causes the air to compress inside the balloon. The deeper the balloon is taken into the water, the greater the pressure, and therefore, the smaller the balloon will become. The pressure from the water will compress the balloon until the air pressure within the balloon can supply an equal amount of force against the water.

Why do **closed containers** sometimes **dent or even collapse** on cool days?

Just as a balloon can become smaller when placed under water, a sealed container can change its shape and even collapse under certain atmospheric conditions. For example, a container that stores gasoline for a lawnmower is usually sealed when it's not being used. If the container were sealed on a warm day, when there was little atmospheric pressure, subsequently on a cool, high-pressure day, the gasoline container would appear crushed.

Since the warmer air inside the container has low pressure and the cooler air outside exerts a larger pressure on the container, the gasoline container would have collapsed a bit. If one opens a can in such a state, a "swoosh" is heard as the high-pressure air rushes into the container to establish an equilibrium on both sides of the container.

Why do some **athletes** go to **high elevations** to train?

Runners have always trekked to the mountains of Colorado to train in higher elevations because of the lower atmospheric pressure that the mountains provide. Since the air is not as dense as it is at lower elevations, the lungs need to work harder to supply the body with a sufficient amount of oxygen. Many athletes feel that training in such conditions gets their bodies used to lower amounts of oxygen. Therefore, when running in a competition at lower elevations, they can compete quite well because their bodies are used to working hard to get a great deal of oxygen.

BUOYANCY

What does it mean to be **buoyant**?

The word buoyant simply the ability to float. Buoyancy occurs whenever the weight of an object is equal to the force of a fluid pushing it upward, and can take place either on or below the surface of a fluid. Whenever a piece of wood is placed in water, it falls, due to gravity, until the buoyant force of the water equals the weight of the wood. When the two equal, the wood floats.

Floating usually refers to a liquid such as water, but can anything **float in a gas**?

Buoyancy occurs when there is a pressure difference within a liquid (such as water) or a gas (such as air). A low-pressure hot air balloon will rise until the weight of the balloon and basket equals the buoyant force of the gas inside the balloon.

Why does a small clump of steel sink, while a 50,000-ton steel ship can float?

In order to remain afloat, a ship needs to displace an amount of fluid equal to its own weight. Therefore, if a clump of steel is placed in water, it will sink because its size wouldn't allow it to displace an amount of water equal to its own weight. In this case, there is no way that the water could apply enough upward force to keep it afloat. A 50,000-ton steel ship can easily stay afloat as long as it can displace 50,000 tons of water. It can do this by widening the hull of the ship and increasing its volume.

What major discovery did **Archimedes** make when he stepped into a bath in the third century B.C.?

To his astonishment, when Archimedes stepped into a tub of water, the water rose! Of course, this wasn't the first time water rose when Archimedes sat in the tub, but it was the first time he would consider the reasons why. He proceeded to conduct an experiment with gold and silver crowns that he immersed in the tub, measuring the water overflowing from the container. Upon the conclusion of his great experiment, Archimedes is said to have run all about his hometown of Syracuse, Sicily, shouting "Eureka!"

Archimedes had discovered a principle of hydrostatics (liquids at rest) that would one day carry his name, Archimedes' Principle. It states that an object immersed in a fluid will experience a buoyant force equal to the weight of the displaced fluid.

What happens to the **buoyancy of the ship** when cargo and passengers are added?

Ship builders must always consider the floatation level of the ship when cargo and passengers are added to it. This increases the weight of the ship. The ship will float as long as the total weight of ship plus contents is less than the weight of the water it displaces. When the weight of the ship and contents exceeds this value, the ship sinks.

The amount the ship lowers in the water as a result of cargo and passengers can be critical for navigation and maneuverability. Large cargo and cruise ships have numbers on the bow of the ship that indicate how far the ship is submerged. If the ship has a twenty-foot draft and the water is only eighteen feet deep, cargo and passengers must be unloaded to allow the ship to rise.

How much **water** does a ship need in order to **float**?

For a ship to stay afloat, it does not need a lot of water, but it needs to be able to displace enough water to equal its weight. Therefore, if a ship were to enter a canal that was just a bit larger than the shape of the ship's hull, it would float just fine as long as there was a small film of water around the entire hull of the ship, as long as the ship were sailing in nonturbulent waters.

How do **hippopotamuses** sink to the bottom of a riverbed?

Hippopotamuses spend over half their day in the water. In order to eat, the hippo, which can reach almost ten feet in length and weigh 10,000 pounds, must sink to feed off the vegetation that grows on the bottom of rivers. However, the hippo has one major problem: his low density forces him to float at the surface, and he is not agile enough to quickly dive down to the bottom and come back up again. In order to reach the bottom, he needs to increase his density, so the buoyant force cannot supply a large enough force to keep the animal afloat. To do this, the hippo exhales, reducing the air in his body to increase his density.

Once he sinks to the bottom, the hippo can feed on the vegetation; however, he cannot inhale to float back up to the surface, so instead the hippopotamus walks up the river's edge or pushes upward off the bottom and emerges at the surface once again.

The Goodyear blimp.

How does the **Goodyear blimp** remain at a particular altitude?

The Goodyear blimp, or "dirigible" as it technically should be called, is a non-rigid airship that floats in the air solely due to the buoyant gas within the giant balloon-like bag. It typically carries over 5,000 cubic meters of helium at a density about seven times less than air. An airship floats in the air in the same manner as a ship floating in water. The weight of the airship must equal the buoyant force of the gas inside the bag. In order to increase the altitude of the blimp, the pilot increases its buoyancy by adding gas from pressurized tanks to the blimp, which expands the flotation bladders, displacing the heavier air and increasing the buoyant force. To lower the airship, the buoyancy is decreased by releasing gas from the flotation bladders, which then decrease in size, displacing less of the heavier air.

Why is **helium** used inside airships instead of **hydrogen**? Isn't hydrogen more buoyant?

Although hydrogen is twice as buoyant as helium, and would be more effective in lifting an airship off the ground, hydrogen gas is extremely

The Hindenburg, a German-owned dirigible, bursts into flames on May 6, 1937, in Lakehurst, New Jersey.

dangerous. In fact, the German-owned Hindenburg, the world's largest airship at the time, was destroyed on May 6, 1937, in Lakehurst, New Jersey, when it exploded into a huge fireball while attempting to land. Thirty-six people died in the explosion.

In 1937, the U.S. was about the only source of helium in the world, mostly from one gas well in Texas. The Nazis wanted to buy helium for their zeppelins, but the U.S. refused to sell it to them—as it was considered a "strategic" resource.

What are **airships** used for?

Since the first airship flight in 1852, by Henry Giffard in France, the dirigible or airship was used predominantly for military purposes. In the early to mid 1900s, airships were used for bombing and surveillance on both sides of the Atlantic. Commercial passenger transportation on airships was conducted for only a few years, while today's modern airships, such as the Goodyear blimp, are used for advertising and for filming sporting events from high elevations.

What happens to the **helium balloon** that a child releases?

If the balloon is tied tightly, the balloon will expand as it increases its altitude. This expansion is caused by the lower atmospheric pressure at higher altitudes. Eventually, the helium's volume increases so much that the rubber balloon breaks, allowing the helium to mix with the outside air.

HYDRAULICS AND PNEUMATICS

What important contribution to fluids did **Blaise Pascal** make in 1647?

Known as Pascal's Principle, the law that Blaise Pascal developed states that any force applied to an enclosed fluid is transmitted in all directions to the walls of the container. This principle is extremely important in the field of hydrostatics and in the development of hydraulics. For example, if a piston pushes against the liquid in a closed cylinder, the force applied by the piston will translate into pressure on the walls of the cylinder. This occurs because liquids cannot be compressed as gases can.

What is **hydraulics**?

Hydraulics is the use of a liquid that, when moved from one place to another, accomplishes by its motion some type of function. The liquid used in hydraulic mechanisms is usually water or oil. Hydraulic engineers design such things as pumps, lifts, faucets, cranes, shock absorbers, and many other devices.

How does a **hydraulic lift** work?

The basic theory behind a hydraulic lift is to multiply forces and give a device mechanical advantage. An automobile lift, used in many automotive repair shops, allows the operator to use very little force to lift an automobile off the ground, by pushing liquid from a small-diameter cylinder and piston through a thin tube that expands into a larger-diameter cylin-

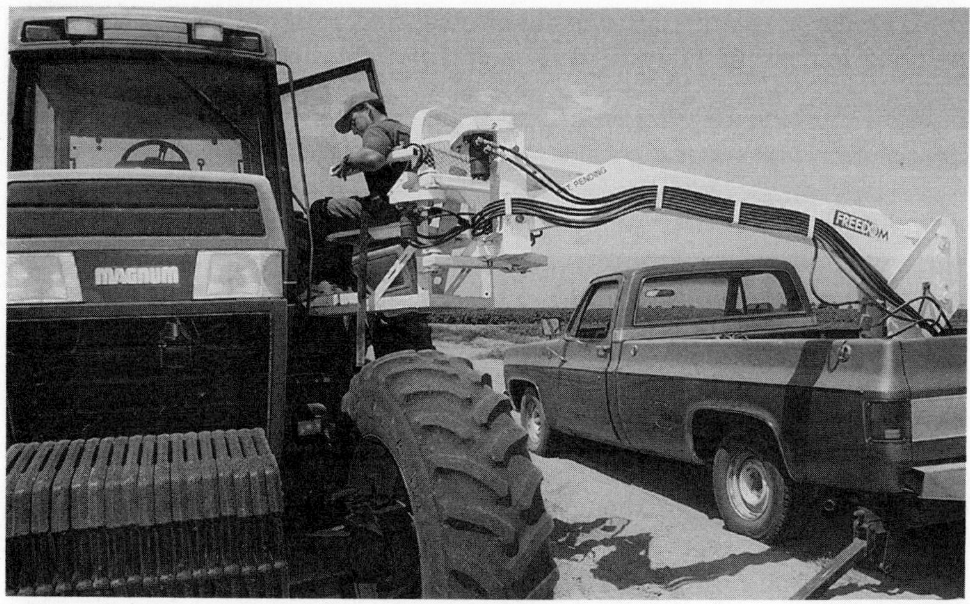

A paraplegic farmer uses a hydraulic lift to help him into his tractor from his pickup truck.

der and piston, which is located beneath the vehicle to be lifted. Since the liquid cannot be compressed like air, the liquid from the small cylinder is pushed into the large cylinder, forcing the large piston to move upward. Although this is a very simplistic view of how a hydraulic lift works, Pascal's Principle states that if a small-area piston pushes a large-area piston, the mechanical advantage can be quite great.

What are some other places where **hydraulic lifts** are used?

Besides their valuable use in auto-repair shops, hydraulic lifts are used in elevating crane and backhoe arms, adjusting flaps on airplanes, and applying brakes in automobiles. It is the non-compressible characteristics of liquids that make hydraulic devices so useful.

What is **pneumatics**?

Whereas hydraulics uses liquids to achieve mechanical advantage, pneumatics uses compressed gas. Since gases can be compressed and stored under pressure, releasing compressed air can provide large forces and torque for machines such as pneumatic drills, hammers, wrenches, and jackhammers.

FLUID DYNAMICS

What is **fluid dynamics**?

Fluid dynamics is the study of fluids in motion. There are several different types of fluid motion: steady flow, where the liquid or gas moves in a constant and predictable manner; unsteady flow, where the fluid makes turns and changes its velocity; and turbulent flow, where the fluid motion is extremely difficult to predict.

What makes a fluid **flow**?

As in all of physics, objects move as a result of forces. Just as a basketball dropped in the air falls to the ground because of a gravitational force, a fluid flows because there is an unbalanced force acting on the liquid—that is, a difference in pressure between two points; fluid will flow in the direction of decreasing pressure.

Why does a river's **current run faster** when the river is **narrow**?

When water flows down a river, the current represents the amount of water that passes by a section of the river in a unit of time. For example, if the current of a river is 2000 l^3/min (cubic liters per minute), this means that, assuming the slope of the river is constant, every minute 2000 cubic liters pass by every section of the river. If a section of the river narrows, the 2000 cubic liters of water still must pass in one minute because the water from behind does not let up in its desire to flow downriver. Since the river is narrower, the water needs to speed up in order to accomplish this task. The principle behind this phenomenon is called continuity.

Why does it always seem windier in the city?

The explanation behind this question is not meteorological, but physical. In major cities, there are skyscrapers and other tall buildings that obstruct the flow of wind. In order to flow past these large obstacles, the wind speed increases in the corridors of the streets and avenues. The same effect can also be found in tunnels and outdoor "breezeways." It is the continuity of the fluid speed rushing through the narrow corridors of streets and avenues that makes the city such a windy place.

AERODYNAMICS

What is **aerodynamics**?

Aerodynamics is a subfield of fluid dynamics dealing specifically with the movement of air and other gases. Engineers studying aerodynamics analyze the flow of gases over and through automobiles, airplanes, golf balls, and other objects that move through air.

What is **Bernoulli's Principle**?

In 1738, a Swiss physicist and mathematician named Daniel Bernoulli discovered that when the speed of a moving fluid increases, such as the wind blowing through the corridors of a city, the pressure of that fluid decreases. Bernoulli discovered this while measuring the pressure of water as it flowed through pipes of different diameters. He found that the speed of the water increased as the diameter of the tube decreased, and that the pressure exerted by the water on the walls of the pipes was less as well. This discovery would prove to be one of the most influential discoveries in fluid mechanics.

How does an airplane wing create "**lift**"?

The wing of an airplane is designed to split the air that approaches the front section of its wing. Part of the air passes under the wing, which is flat on the bottom, while the rest passes over the upper section, which is curved. The curved upper section makes the air above the wing travel a greater distance than the air under the wing. As a result, and due to the continuity of flow, the air above the wing needs to travel faster than the air underneath the wing. According to Bernoulli, if the air travels faster on top, it must also have less pressure than the bottom section. The resulting pressure difference produces the lift needed to keep the plane airborne.

What is **drag**?

Drag is a force that tries to slow down an object in air flow. An object is often said to be "aerodynamic" when its drag forces are kept to a minimum.

There are two types of drag: parasitic and induced. Parasitic drag is the friction that a fluid experiences when it comes in contact with a moving airplane wing, automobile, or any other object. The amount of drag also depends upon the physical properties of the fluid, such as viscosity. The more viscous it is, the thicker and slower the fluid. Another variable that affects drag is the shape of the object passing through the fluid. A wide, rectangular barge will have more drag through the water than a V-shaped cigarette speedboat. It is a combination of the viscosity of the fluid and the shape of the object that determines parasitic drag.

Induced drag is a consequence of the lift generated by the wing. It is a function of the angle of attack of the wing—the lower the angle of attack, the smaller the induced drag.

What are **streamlines**?

Streamlines are lines that represent the flow of a fluid around an object or through another fluid, when the path is repeatable—that is, during non-turbulent flow. Streamlines are used predominantly in wind tunnel testing of airfoils (wings) and automobiles. A windtunnel is a closed

Streamline testing of a Ferrari in a windtunnel.

chamber with vents in the front and rear of the tunnel that allow wind and streams of smoke called streamlines to pass around an object. If the streamline smoke appears to be flowing in a gentle pattern without breaking up, then the object is considered aerodynamic. If the smoke breaks up upon encountering sections of an object, that section may not be aerodynamically smooth.

Why do **golf balls** have **dimples**?

Golfing has been around for several centuries, but dimpled golf balls have only been around for less than a hundred years. The dimples in golf balls, first introduced by the Spalding Company in 1908, can double the distance a golf ball can fly. The dimples actually drag a thin layer of air completely around the golf ball. When hit with a slight backspin, the wind passing over the top section of a dimpled ball flows in the same direction as the rest of the air around the ball. This establishes an area of low pressure above the ball. The air is carried along the surface of the ball until it strikes the underside of the ball, where it meets the wind head on. This lowers the velocity and creates an area of

The dimples on golf balls can double the distance that a golf ball can fly.

high pressure under the ball. According to Bernoulli's Principle, such a pressure difference should provide a lifting force on the ball. Therefore, since the dimples cause air to flow around the ball, they create a difference in pressure, which results in more lift, which affords the ball a few more seconds worth of flight.

How does a **curve ball** curve?

The curve ball employs Bernoulli's Principle just as an airplane wing or discus employs lift to help them fly. However, instead of using lift to fly, the curve ball uses the pressure difference to move the ball sideways. Just as a golf ball's dimples enable a thin layer of air to travel around the ball, the stitches on a baseball do the same thing. When the pitcher gives the ball a spin, the layer of air traveling around one side of the ball moves in the direction of the spin and with the airflow, while on the other side of the ball, the air next to the ball moves against the airflow. This difference in airflow creates different pressures that translate into a sideways lift, called a deflection force. It is this force that causes the ball to curve to one side and deceive the batter.

What is the most **aerodynamic shape**?

Some think the narrower and more needle-like an object is, the lower its drag force will be. Although a needle-head cuts easily through the wind, the problem emerges at the tail end, where the wind becomes turbulent and forms small eddy currents that hinder the streamline flow of air. The optimum shape depends on the velocity of the object.

For speeds lower than the speed of sound, the most aerodynamically efficient shape is the teardrop. The teardrop has a rounded nose that tapers as it moves backward, forming a narrow, yet rounded tail, which gradually brings the air around the object back together instead of creating eddy currents.

Why is it better for a discus to be thrown into the wind instead of with the wind?

In most sports, throwing or traveling with the wind at your back (this is called tail wind) is a lot easier than working against the wind (head wind). In football, teams flip a coin to determine who will travel with the wind. In sailing, it is easier and faster to travel perpendicular to or with the wind and more tedious to travel against the wind. In track, the world record in the 100-meter dash can easily be broken if running with a strong tail wind. In most sports, having the wind at your back can be a major advantage.

However, in the field event of discus throwing, the advantage comes when there is a head wind. In fact, it has been documented that a discus can travel up to eight meters farther while experiencing a head wind of only 10 m/s (meters per second). Although the discus still experiences a drag force from the head wind, the lift that the discus gets from pressure differences over and under the disc is substantially more significant than the drag force. Because the discus will remain in the air longer, it will travel farther.

At high velocities, such as a jet airplane or a bullet may travel, other shapes are better. For turbulent flow, the least drag comes from having a blunt end, which intentionally causes turbulence. The rest of the air then flows smoothly over the region of turbulence behind the object.

What determines the difference between **laminar and turbulent velocities**?

Eddy currents can form when a fluid moves from low, laminar velocities, to fast, turbulent velocities. The point at which the flow of a fluid changes from laminar to turbulent depends upon several variables and

What are microbursts?

Microbursts or downbursts are caused by the evaporation of rain when water droplets fall from a thunderstorm cloud. The evaporation of the precipitation quickly cools the air, which makes it heavier than the warmer air in the rest of the cloud. The cold air then plummets to the ground and shoots out in all directions. The wind from the microbursts can reach speeds of over 100 mph (miles per hour).

is represented by the Reynolds number. Those variables are the velocity, fluid density, cross-sectional area, and the viscosity of the fluid.

What is an example of a transition from **laminar flow** (no turbulence) to **turbulent flow**?

The transition to a turbulent flow occurs mainly when the velocity of a fluid increases. An example of the transition and the ensuing eddy currents can be seen when smoke rises from a cigarette. When smoke emerges from a cigarette, the smoke fluid rises slowly, but within approximately two to three centimeters from the burning tobacco, the smoke speeds up because of its buoyancy around the cooler, surrounding air. This is where the transformation from laminar to turbulent flow takes place, and where the turbulence forms unpredictable eddy currents in the smoke.

Why are microbursts so **dangerous** to airplanes?

When a plane approaches the front of a microburst, the wind is extremely fast, increasing the speed over the wings, while decreasing the pressure, and lifting the plane higher into the air. When the plane hits the

The Wright brothers prepare for the take off of the first powered flight, at Kitty Hawk, North Carolina, on December 17, 1903.

down draft, it plummets toward the ground until it emerges from the burst. Finally, upon escaping the burst, the plane experiences a massive tailwind, which reduces the wind velocity passing over the wings, and as a result, reduces the lift and maneuverability of the airplane.

From ground level, many people think that microbursts are actually tornadoes, for they carry high winds, are difficult to predict, and make a loud sound. Next to pilot error, microbursts are the second-highest cause of airplane accidents. In fact, over thirty planes have crashed as a result of microbursts. The burst may only be a scary ride when traveling at high altitudes, for the plane would have to fall extremely far before crashing. However, if occurring at lower altitudes, microbursts can easily push planes down until they lose control and hit the ground.

What happened at **Kitty Hawk,** North Carolina, on December 17th, 1903?

It was on this date that brothers Orville and Wilbur Wright warmed up the engines on their Wright 1903 Flyer and took off into the blustery winter air. On its first flight, Orville flew the Flyer for a total of 12 seconds and traveled a distance of 120 feet. Later that cold winter day, Wilbur flew for nearly a minute and traveled 852 feet. The Wright 1903 Flyer, which weighed only 600 pounds and had a wingspan of 40 feet, made only four runs that day, for after Wilbur's 852-foot flight, the wind tossed the plane end over end, breaking the wings, engine, and chain guides.

How are **airplane controls** different from the controls in an automobile?

Automobiles travel on two-dimensional surfaces, and therefore only need two separate controls: the accelerator and brake to control the forward movement, and the steering wheel to control side-to-side movements. Airplanes, on the other hand, travel in three-dimensional space. The forward thrust on an airplane is controlled by the throttle, and the "braking" is achieved by closing the throttle and increasing the drag, usually by deploying the plane's flaps. Note that a plane cannot travel in reverse, as a car can. Yaw, which is responsible for the side-to-side movement of a plane, is controlled by the plane's rudder.

To control the pitch, or up-and-down orientation of the nose, the pilot uses elevators or horizontal control surfaces near the plane's rudder. To roll the plane (the rotation of the plane about an axis that goes from the nose to the tail), the pilot uses control surfaces on the back end of the wings called ailerons.

THE SOUND BARRIER

What is a **shock wave**?

Just as a boat moving through the water forms a series of V-shaped waves, airplanes create conical (cone-shaped) waves as they fly through the air. The waves that the airplane produces are waves of compressed air. When an aircraft reaches the speed of sound, Mach 1, the plane's pressure waves are so compressed that the sound waves overlap each other, creating a shock wave. Otherwise known as a sonic boom, the

Chuck Yeager beside the Bell X-1 after the first powered take off of a supersonic plane.

shock wave creates one single, loud sonic boom heard by observers on the ground. When the plane travels slower than the speed of sound, the sound waves do not overlap and instead of hearing a sonic boom, observers simply hear the delayed sound of the plane.

If **Mach 1** is the speed of sound, what is Mach 2?

Mach is the ratio of a velocity to the speed of sound, so Mach 2 is two times the speed of sound, Mach 3.5 is three and a half times the speed of sound, etc. Any velocity greater than Mach 1 is referred to as "supersonic."

Who was the first pilot to **break the sound barrier**?

On October 14, 1947, Chuck Yeager broke the sound barrier in his Bell X-1 test plane, "Glamorous Glennis." In order to reach the sound barrier, the X-1 was carried in the belly of a B-29 bomber to an altitude of 12,000 feet where it was dropped. The X-1's rocket engine ignited and

Yeager took the plane to an altitude of 43,000 feet. At this altitude, Yeager was able to break the sound barrier by traveling 660 mph (miles per hour). The X-1 experienced a turbulent set of compression waves just before he broke past the barrier at Mach 1.05. Yeager kept the plane at this supersonic speed for a few moments before he cut off the rocket engine and headed back toward Earth.

Why did ChuckYeager go to such a **high altitude** to break the sound barrier?

Sound travels approximately 760 mph (miles per hour) in warm, dense air as found close to sea level. The cooler and less-dense air is, however, the lower the speed of sound. Since air is less dense at higher elevations, physicists and engineers felt it would be easier to break the sound barrier at those elevations. Knowing the temperature and density of the air at 40,000 feet above sea level, scientists determined that the speed of sound would be reduced to only 660 miles per hour. As an added bonus, engineers found that not only was the speed of sound slower at such elevations, but when the air has such low density, the parasitic drag (the drag due to friction), is very low as well. Therefore, to break the sound barrier, Yeager traveled as far up as 43,000 feet above sea level to both reduce the sound barrier and decrease the parasitic drag.

What **concerns** did pilots and engineers have about breaking the sound barrier?

To reach the sound barrier in an airplane was a major goal for many in the aeronautical field, a goal that carried some uncertainties. Pilots and engineers alike wondered and feared what would happen to a plane's maneuverability when it broke through the pressure wave created by the plane's own forward thrust, as well as what would happen to the plane itself, structurally.

Near the end of the Second World War, there where several very powerful models of fighter planes. These planes were very strong, and had powerful engines and experienced pilots. A number of pilots died when their planes broke apart in mid-air, often when in dives. There where two problems with these aircraft: first, the wings were not swept back, and second, they where driven by propellers. As the shock wave forms near Mach 1, it bends backward from the nose of the plane, like a bow wave on a boat. If the shock wave encounters the wings (that is, the wing extends through the shock front), there are tremendous forces on the wings. In a supersonic plane the wing is always designed to be fully behind the shock front, because the shock front can tear the wing off the plane. The propeller causes a pulsation in the pressure on the wing: every time one of the blades goes by, it produces a region of slightly higher pressure behind it, followed by a region of low pressure. All of these things came together and helped cause mid-air structural failures of the WWII fighter planes.

SUPERSONIC FLIGHT

Why are the **angles of wings** important for supersonic flight?

When a plane breaks the sound barrier, the air in front of the plane has a difficult time moving out of the plane's way, and the compression of the displaced air creates a shock wave. In order to break the sound barrier with less difficulty, aeronautical engineers have designed more aerodynamic fuselages and efficient wing designs. As mentioned above, wings on supersonic planes must remain behind the shock front to prevent structural failure and allow the plane to maneuver safely. The swept-back wing design, as found on many commercial airplanes, allows the airplane to accelerate easily and faster before major pressure builds up around the wings. The delta wings, as found on many jet fighters, are large and extremely thin, to keep the wings behind the shock front while increasing lift and reducing drag.

Problems can also occur when using swept-back wings. As a plane moves faster, the center of lift on the wings can move too far backward, causing unbalanced forces on the plane, which can affect the maneuverability and safety of the plane. One such plane is the well-known executive plane called the Lear Jet. Lears are notorious for this problem, and are therefore limited to speeds of about Mach 0.8.

What is the fastest aircraft?

Just like Chuck Yeager's Bell X-1, which was the first plane to break the sound barrier, the X-15A-2, the fastest aircraft ever built, was dropped from the belly of a B-52 bomber. When released, the X-15A-2's rockets ignited, taking it to a maximum speed of 4,534 mph (miles per hour). That speed, which is equivalent to 7,297 km/h (kilometers per hour), flew at 6 times the speed of sound. The SR-71 (blackbird) is the fastest known aircraft that can take off under its own power.

How fast can the **Concorde** fly?

Since 1973, the Concorde has been a symbol of fast and expensive air travel for business people. The plane, which is flown by British Airways and Air France to only 85 different destinations, is an inefficient but fast plane. The sleek, delta wing design and pinpoint nose, which tilts down during liftoff and landings, can achieve speeds of up to Mach 2.2 at 50,000 feet above sea level.

What does the future hold for **air travel**?

The planes from Boeing, Airbus, and McDonald-Douglas have been the staples for airplane travel. However, in the future, airplanes may look quite different than they do today. Aeronautical engineers have made preliminary designs of "flying wings," which look more like stealth bombers since they eliminate the fuselage of a conventional plane. Within these "wings" might sit 600 to 800 passengers. Many people are critical of this idea: first, the airline industry sees little, if any, demand for planes that accommodate that many people, and airports feel that redesigning terminals and runways would be necessary to accommodate such "wings." If there are any changes on the horizon, many experts feel they will simply be improvements—to the jet engines, cockpit controls, fuselage, and wing materials—on the already successful designs of today's modern commercial fleets.

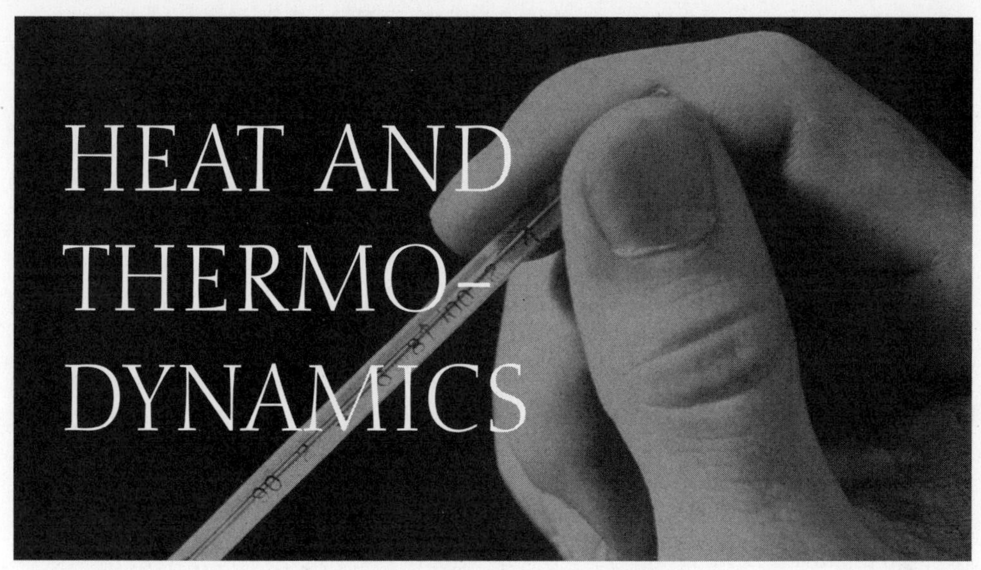

HEAT

MEASUREMENT

What is the difference between **heat** and **temperature**?

Heat is the amount of internal kinetic energy of atoms and molecules that flows from a warmer to a cooler environment in an effort to reach equilibrium. For example, if your finger touches a hot bagel, the thermal energy flows from the bagel to your finger in an attempt to equalize the temperature between the bagel and your finger.

Temperature is a scale for measuring thermal energy by showing how warm or cold an object is relative to something else. Typical temperature scales, such as Celsius and Fahrenheit, were created by scientists to measure hotness or coldness relative to the freezing and boiling points of water.

What is a **caloric**?

In earlier times, heat was thought to be something called caloric. If one had a lot of caloric, then that person was warm; if the person did not have as much caloric, the person was cold. Even in the early stages of

What is a calorie?

Heat is a form of energy, and therefore uses the unit named after James Prescott Joule. Although the joule is the international standard for measuring energy, heat can be measured in calories. A calorie defines the amount of heat needed to increase one gram of water by one degree Celsius or Kelvin. The energy required for one calorie is 4.186 joules, which is a relatively small amount of energy. Nutritionists also use the term "calorie," to describe the amount of energy a particular food can provide a consumer. However, a nutritional calorie is really a kilocalorie (1000 calories), which is also rendered as the capitalized "Calorie."

Another unit used to measure heat is the British Thermal Unit, or Btu. The Btu is similar to the calorie in that it is the amount of energy needed to increase the temperature of one pound of water by one degree Fahrenheit. This unit, which is used only by countries such as the U.S. that still employ the English standard method of measurement, is equal to 252 calories.

studying heat, scientists knew it flowed and was something that could be gained by warmer objects. It was not until the late eighteenth to mid-nineteenth century that physicists determined that heat was not a material object such as a caloric, but simply the kinetic, thermal energy of the atoms within objects, which could easily be transferred to other objects.

What is the **Fahrenheit** temperature scale?

Temperature scales are, in a sense, artificial; they are created by humans. German physicist Gabriel Fahrenheit developed the first well-known temperature scale in 1714. Equipped with the first mercury thermometer, Fahrenheit defined a scale in which the freezing point of water was 32°F and the boiling point was 212°F.

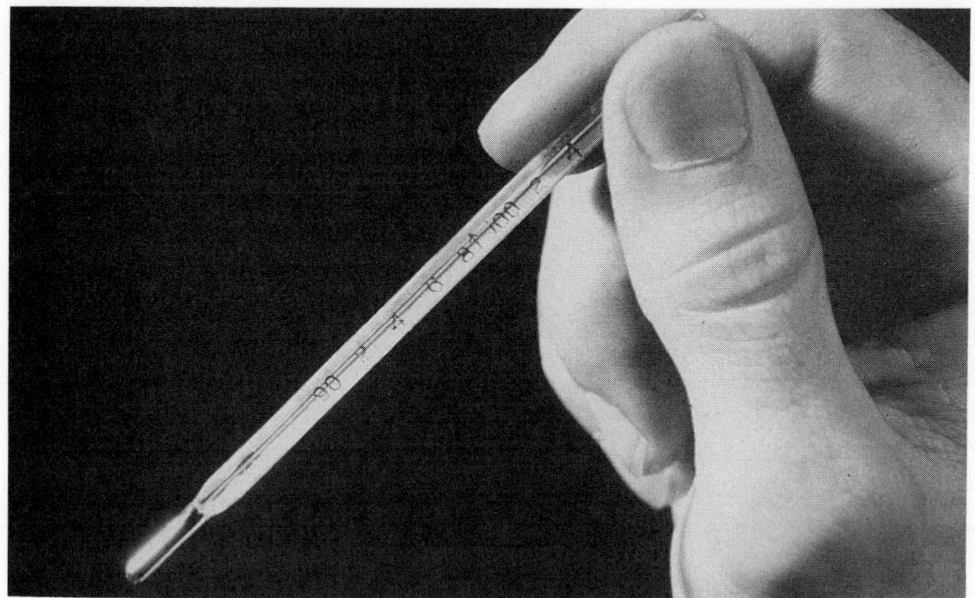

A medical thermometer.

Why did Gabriel Fahrenheit define the **freezing point of water** to be 32°F instead of 0?

Fahrenheit did not define 32° as the freezing point of water. Instead, he defined 0° as the freezing point of a water and salt mixture. Since salt lowers the freezing point of water, the freezing point for this mixture was lower than it would have been for plain water. Upon defining the degree intervals between the freezing and boiling points of the water and salt mixture, and he found that water itself freezes at 32°F.

How do **thermometers** work?

Most materials expand as they gain heat energy; thermometers meant for the average consumer use this principle to measure temperature. The thermometer's hollowed-out body and bulb contain a predetermined amount of alcohol or mercury that, when cool, remains in the bulb. But when the temperature rises, the liquid expands, forcing its way up the narrow capillary tube in the thermometer. As the liquid rises, an adjacent scale indicates the actual temperature reading.

Mercury droplets.

Why is **mercury** used in some thermometers?

Fahrenheit was the first person to use mercury instead of alcohol in a thermometer. The reason why Fahrenheit, and later other scientists, used mercury was because it had a constant and significant expansion rate as its temperature increases. In other words, for every degree the mercury's temperature increases, the difference in which the mercury expands or "rises" in a thermometer is fairly noticeable and in equal amounts; therefore, the degrees on the Fahrenheit scale are evenly and widely spaced, making developing, manufacturing, and reading thermometers relatively easy.

Alcohol thermometers are, however, often used in extremely cold climates—say, Alaska or Siberia—because mercury will freeze.

If a mercury thermometer breaks, is it **dangerous** to touch the **spilled mercury**?

Mercury is a dangerous metal that can cause great damage, especially to the kidneys and nervous system. Mercury from a broken thermometer should not be touched, but instead scooped up and disposed of as a hazardous substance. Although mercury poisoning will not occur unless larger doses are ingested or come in contact with one's skin, one should take proper precautionary measures when handling it anyway. Mercury is not only found in thermometers but also in barometers, which measure atmospheric pressure.

Who invented the **first thermometer**?

Although Galileo is credited with developing the first thermometer in 1592, it was not until 1713 that Gabriel Fahrenheit developed the first closed-tube mercury thermometer. Combined with the temperature scale he defined the following year, Fahrenheit made a significant contribution to science.

Can **electricity** be used to measure temperature?

One of the most common devices for measuring temperature in scientific laboratories employs electricity. Such a device uses two different metal wires joined together at two separate "junctions." This device, usually made of copper and a copper-nickel alloy, is called a thermocouple. When the temperature of an object changes, the voltage difference on the thermocouple changes. The difference is measured and converted into a temperature reading. The thermocouple has the ability to read a wide range of temperatures from -270°C to 2300°C. It is a rather complicated device, which is why it is used primarily in laboratory environments.

How can a single **piece of metal** measure the temperature of objects?

A thermistor is a temperature-sensitive resistor; it determines the temperature of an object by measuring how difficult it is to send electrical current through a piece of metal. Thermistors are usually made of nickel, manganese, and cobalt, or for higher temperatures, platinum. When the thermistor is placed on or in an object that needs to be measured, its electrical resistance indicates the temperature of the object.

Can thermal energy or heat be seen through **cameras**?

When an object is hot, it releases thermal energy known as heat. Heat produces infrared electromagnetic waves and can be detected by infrared cameras. These cameras are often used to determine the amount of heat radiating from a body or object. When attempting to measure the temperature over a large area, thermometers are not practical. Therefore, scientists use infrared cameras to take "temperature pictures," known to meteorologists as thermographs.

How are **thermographs** used?

Thermographs, which detect the amount of heat emanating from objects or regions, use colors to determine the temperature. Typically,

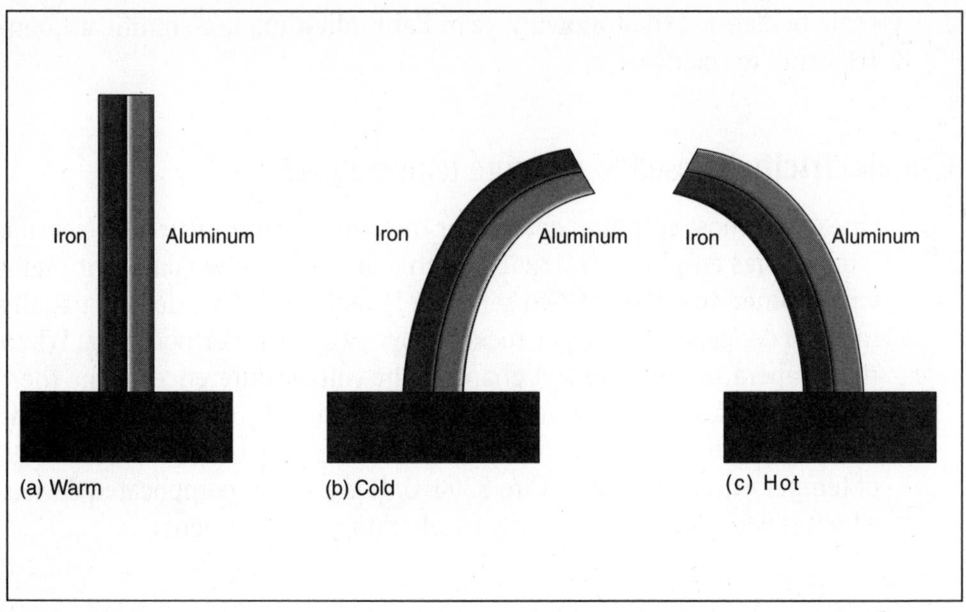

Representation of a thermally induced change in a bimetallic strip, which is used for thermometers and thermostats, among other technology.

red indicates the warmest temperatures, while blue indicates cooler temperatures. Thermographs are used throughout science, but are well noted for their use in identifying cancerous tumors in the body, and measuring the temperature over regions of the Earth.

What is the difference between a **thermometer** and a **thermostat**?

Thermometers measure the heat energy released from an object, whereas a thermostat not only measures the temperature, but also controls heating and cooling systems.

What is a **bimetallic thermometer**?

A bimetallic thermometer uses the expanding properties of two different metals to gauge temperature. In such a thermometer, two metals, typically iron and copper, are welded or coiled together to form a single, bimetallic strip. When the temperature increases, copper responds more dramatically to the change and expands more than the iron, causing the bimetallic strip to bend. The strip has a pointer attached to a scale, which indicates the temperature. Most bimetallic strips are not used in thermometers but in thermostats that regulate the heat of irons, furnaces, and other heating devices.

What is a **bimetallic thermostat**?

The bimetallic thermostat is similar to the bimetallic thermometer. Its main function is to determine if the heat in a building needs to turn on or off. The bimetallic thermostat consists of an electrical circuit and two different metals fused together to form a single strip of metal. When the temperature rises, the metal under the strip expands more than the upper metal, causing the strip to bend upwards. When the strip bends upwards, it loses contact with an electrical wire and cuts off the current to the heater. When the temperature of the bimetallic strip cools, the strip straightens out again and makes contact with the electrical circuit. The flow of electricity sends the message to the heater that it needs to turn on to heat the building.

Who developed the **Celsius** scale?

The Celsius scale is named after a person whose life work was dedicated to astronomy. Anders Celsius, a Swedish astronomer, spent most of his life studying the heavens. Before developing the Celsius temperature scale in 1742, he published a book in 1733 documenting the details of hundreds of observations he had made of the aurora borealis, or northern lights. Celsius died in 1744, at the age of forty-three.

What is the **Kelvin** scale?

The Kelvin temperature scale, developed by Lord William Thomson Kelvin in 1848, is widely used by scientists throughout the world. Absolute zero is a theoretical temperature indicating zero heat energy. Each degree on the Kelvin scale is equal to a degree on the Celsius scale, but the difference is where zero is. For Celsius, 0° is the freezing point of water; for Kelvin, the zero point is at absolute zero. Therefore, 0

What is the temperature scale used by the majority of the world's population?

Just as most people in the world use the metric scale, they also use the Celsius scale for temperature. Developed in the early eighteenth century, the scale is based on the freezing and boiling points of water. Originally, the freezing point of water was deemed to be 100°, while the boiling point was 0°. Swedish biologist Carolus Linnaeus, a colleague of Celsius at Upsalla University in Sweden and known for his classification of plants and animals, switched the Celsius scale so the freezing point of water was 0° and the boiling point was 100° Celsius.

Kelvin is equal to -273.15°Celsius; 0°Celsius is equal to 273.15 Kelvin. Most scientists feel the Kelvin is a better scale because it does not compare the temperature to freezing or boiling water, but instead to the absolute lowest temperature possible.

How are the **Fahrenheit and Rankine** scales related?

Just as the Kelvin and Celsius scales are quite similar to each other, the Rankine and Fahrenheit scales are similar as well. The Rankine scale, like the Kelvin scale, begins at absolute zero. Rankine's zero is equivalent to -459.7° Fahrenheit, and progresses upwards, with each degree equal to a degree on the Fahrenheit scale.

What are the **equivalent temperatures** of some temperature scales?

The following table provides some examples for the four main temperature scales:

Amount of Heat	Celsius	Kelvin	Fahrenheit	Rankine
Absolute zero	-273.2	0	-459.7	0

Amount of Heat	Celsius	Kelvin	Fahrenheit	Rankine
Freezing water	0	-273.2	32	491.7
Typical human body	37	273.2	98.6	558.3
Boiling water	100	373.2	212	671.7

What are some **conversion formulas** for the different temperature scales?

From	To	Formula
Fahrenheit (F)	Celsius (C)	C = 5/9 (F – 32)
Celsius (C)	Fahrenheit (F)	F = 9/5 C + 32
Kelvin (K)	Celsius (C)	C = K – 273.2
Celsius (C)	Kelvin (K)	K = C + 273.2
Fahrenheit (F)	Rankine (R)	F = R – 459.7
Rankine (R)	Fahrenheit (F)	R = F + 459.7

ABSOLUTE ZERO

What is the **lowest possible temperature**?

The lowest possible temperature is called absolute zero (0 Kelvin). It is achieved when the pressure of a gas is lowered to zero and there is no heat. Since the pressure of a gas can never be completely reduced to zero, absolute zero can never be achieved, and thus remains a purely theoretical temperature.

Has anyone ever come close to **absolute zero**?

Although no one has been able to reach absolute zero, physicists have been able to reach temperatures of millikelvins (thousandths of a Kelvin) in the lab. In some applications, microkelvins have been achieved (millionths of a Kelvin). It is pointless to publish a "lowest

How do astronomers determine the temperature of the sun?

When iron is hot, you can feel the heat radiating from it. That heat is in the form of infrared waves leaving the iron. When iron gets extremely hot, it produces a red glow—and when it gets even warmer, it can take on a whitish glow. The temperature of iron and other objects can be measured by the amount of heat flowing from it, but can also be measured by the light it emits.

Scientists measure the temperature of stars and sun by analyzing the color and brightness that the stellar objects produce. From such experiments, physicists have concluded that the surface of the sun is approximately 5500° Celsius (9900° Fahrenheit).

temperature," for the trend has been that every few months a new record is set, getting closer and closer to absolute zero.

Is there an **absolute highest temperature** that can be achieved?

Although there is an absolute zero temperature, where heat and pressure no longer exist, there is no known absolute high temperature. The highest temperatures achieved to date have been from nuclear explosions, where the temperature can reach as high as one hundred million Kelvin.

What are the **average surface temperatures** of the **planets** in our solar system?

For the planets that have atmospheres (mixtures of gases surrounding the surface of a planet), the average temperature stays relatively constant, for the atmosphere acts as a type of insulator. These planets have only small variations in the temperature when a section of the planet faces away from the sun.

Planet	Daytime Temp Range (°Celsius)
Mercury	−173 to 427
Venus	427
Earth	−25 to 35
Mars	−63 to 27
Jupiter	−163 to −123
Saturn	−178
Uranus	−215
Neptune	−217
Pluto	−233

States of Matter

What are the different **states of matter**?

The three major phases of matter are the solid, liquid, and gaseous states. (Some scientists recognize plasma, a state closely related to gas, as a fourth state of matter.) The particular chemical characteristic a material possesses determines when it changes from one phase to another. Solid phases are found at lower temperatures; as the amount of internal energy increases, the material changes from the solid to the liquid and then to the gaseous phase (but rarely to the plasma state). Water, for example, changes from ice, its solid state, to liquid water, and finally to steam, its gaseous state.

Is it possible for substances to **skip the liquid phase** of matter?

It is quite possible for some materials to skip the liquid phase. In fact, carbon dioxide (CO_2) can skip right from the solid stage of dry ice directly to its gaseous state, given the right amount of energy. This process is called sublimation.

Plasma sounds like something out of science fiction. What is it?

The plasma state occurs when the atoms in a gaseous molecule become ionized, or charged particles. When a gas is submitted to extremely high temperatures, often tens of thousands Kelvin, the collisions of the atoms in the gas are so great that they break apart. Since the atoms are literally in pieces, each piece has its own charge, forming ionized gas or plasma.

Plasma is created whenever the temperature is high enough to break atoms apart. One example of a plasma state is in the sun. Temperatures in the plasma core of the sun reach as much as 15 million Kelvin. Here on Earth, lightning bolts produce plasma for a fraction of a second while they light up the sky.

What determines the amount of **energy** required to **increase the temperature** of a substance?

The specific heat capacity determines the amount of energy (measured in joules or calories) needed to increase the temperature of a substance of some mass (measured in grams or kilograms) by one degree Celsius. For example, to raise one gram of water by a degree Celsius requires one calorie of energy. To increase a gram of copper by one degree requires only .09 calories. The following lists specific heat capacities (the energy in joules it would take to raise one kilogram of substance by one degree Celsius) of certain common solids, liquids, and gases:

Material	Specific Heat (joules / kg°C)
Aluminum	899
Copper	387
Glass	837
Gold	129
Lead	129
Iron	445
Wood	1700
Mercury	140
Silver	235
Water	4186
Ice	2090
Steam	2010

Material	Specific Heat (joules / kg°C)
Ammonia	2190
Carbon Dioxide	833
Nitrogen	1040
Oxygen	912

Is the same amount of heat needed to **change ice into water** as is needed to **change water into steam**?

No. Vaporization—changing liquid to gas—requires more energy than fusion—changing a solid to a liquid. The amount of energy needed to change phase is called latent heat. For example, the latent heat of fusion for changing ice into liquid water is 79.7 calories. The latent heat of vaporization to change one gram of water to steam is 541 calories.

Heat Transfer

What are some ways of **transferring heat**?

Whether it's warm air circulating around a living room, a young child walking on hot sand, or an alligator sunning itself in the middle of a golf course, heat is being transferred from one object to another. Circulating warm air transfers heat through convection; conduction is achieved through contact with a warm surface; and exposure to electromagnetic radiation, specifically infrared waves, is called radiation.

How does hot air blowing out of a **heating duct** warm a room?

Convection is the movement of heat through a fluid (such as liquid or gas). When warm air blows into a room, it warms the air directly around it. The warmer air rises because it is lighter than the dense cool air surrounding it; the currents of the warm air move upward and essentially float on the denser, cooler air molecules below. When installing hot-air heating systems, the ducts are usually placed low on walls or on the floor. This way, the rising warm air will pass by the occupants of the room.

How do **convection currents** create sea breezes?

In general, the convection currents in the Earth's atmosphere cause wind. During a day at the shore, the sun warms the air by radiating infrared light. The air above the shore, however, warms up easier than the air over the ocean. The warm air over land rises in convection currents and ocean air flows toward the shore to "fill in the gap" left by the rising warm air. This flow of cooler air from the ocean creates what is known as a sea breeze.

In the evening, when the sun dips below the horizon, the air over the ocean is warmer than the shore air, and the reverse takes place. The warmer ocean air rises while a gentle breeze fills in the gap from the shore.

How does heat flow throughout an iron **frying pan**?

When a cast-iron frying pan is placed on a stove, the heat from the pan will inevitably reach the bare iron handle. The transfer of heat from the pan to the handle is through a process called conduction. Conduction occurs when heat flows throughout a material; in this case, the internal energy is being conducted throughout the entire cast-iron pan. The internal kinetic energy of the iron atoms makes them vibrate back and forth quite violently. Collisions take place, and the internal energy of the pan increases. Although the stove only heats up the pan section of the frying pan, the conductive properties of iron causes the internal energy to spread throughout the entire pan and whatever other objects are touching it.

Good conductors of heat are often metals with loose free electrons. Free electrons refer to electrons that can easily "jump" from one atom to the next when given enough kinetic energy. When the internal energy of a conductor increases, it is the free electrons moving from atom to atom, as well as the collisions taking place between the atoms, that allows the heat to move into the handle.

How do **igloos** keep Eskimos warm?

Although snow and ice are definitely not sources of heat, air pockets within snow and ice function as excellent insulators of heat. Many small

Why does it feel cold on a tile floor and not as cold on a rug?

The transfer of heat occurs when there is a difference in temperature between two objects. Heat can only be moved from a warm to a cooler object. And the larger the temperature difference between two objects, the greater the amount of heat that needs to be transferred.

A tile floor feels cooler than a carpeted floor because tile is a better conductor, that is, mover, of heat. Both materials are actually the same temperature; however, since tile is a better conductor of heat than carpeting, it accepts heat faster from your body than the carpeting does. If your foot loses heat faster from the tile, then your foot will feel cooler on the tile than on the carpeting. This is another example of heat flowing by means of conduction.

mammals build snow dens to keep themselves warm, thereby taking advantage of the insulating properties of the snow.

Many farmers use the same principle to protect their crops during sub-zero temperatures. They spray water on the crops, and when the water freezes, the plants are insulated by the poor conductive properties of the ice.

How does the sun transfer energy by **radiation**?

When an alligator suns itself on a golfing green, it's absorbing energy directly from the sun. This energy emitted from our closest star is called radiation. Instead of transferring heat through convection or conduction, heat radiation travels by means of infrared light waves. When infrared light waves strike an alligator (or a person or a plant), the energy of the infrared waves excites the object's molecules and causes them to vibrate. It is the vibrations caused by the infrared radiation that warms the alligator.

Air pockets within snow and ice function as excellent insulators of heat.

Why is it often **cooler** to wear **white clothes,** rather than black clothes?

White surfaces reflect all the colors of the rainbow, whereas black absorbs all the colors in the light spectrum. The absorption of this energy causes the atoms inside the object to get excited, vibrate, and increase its internal energy. When internal energy increases, the temperature of the object increases as well. Because black-colored materials and objects absorb more energy than lighter-colored materials, they get hotter.

THERMODYNAMICS

What is **thermodynamics**?

Thermodynamics is the field of physics that studies the movement of heat. Specifically, physicists who study thermodynamics observe how heat can be used and transformed into different forms of useable energy. There are a zeroth, first, second, and third law of thermodynamics.

THE ZEROTH LAW

What is the **zeroth law** of thermodynamics?

The zeroth law is a very simple law, and for that reason many physicists do not consider it to be important. The zeroth law states that temperature is a method of determining if heat will flow from one object to another. If two objects are the same temperature, then they will exchange no heat between each other. However, if one object has a higher temperature, it will give up some of its internal energy, in the form of heat, to the other object, until thermal equilibrium is achieved.

THE FIRST LAW

What important contribution to thermodynamics did James Prescott **Joule** make?

James Joule made a significant contribution to science when, in an experiment, he determined that revolving paddles warmed up water inside a container. Joule's experiment helped form the first law of thermodynamics. The mechanical energy of the paddles turning in the container caused the temperature of the water to rise. He proved that work and energy increases the temperature of the surroundings. This discovery by Joule led to the first law of thermodynamics.

What is the **first law** of thermodynamics?

The first law of thermodynamics is a restatement of the conservation of energy. It says that energy can change form, but it must be conserved.

How do clouds form?

As warm air rises into the atmosphere through convection currents, the air expands as it experiences less atmospheric pressure. During the expansion, the warm water vapor quickly cools and condenses, forming water droplets in the air. When the droplets begin to accumulate, they attach themselves to other particles in the air and form clouds.

Heat is energy and can be changed into different forms of energy—be it mechanical or electrical or any other form.

A steam turbine is an excellent example of the use of the first law of thermodynamics. The steam, a form of energy, gives up some of its energy to turn the turbines, which in turn transfers some of its energy to generate electricity or some other type of mechanical movement. Through this process, however, the total amount of energy is fixed. It is only transferred from one form to another.

THE SECOND LAW

What is the **second law** of thermodynamics?

The second law of thermodynamics has two parts. The first is similar to the zeroth law of thermodynamics in that it states that heat will only flow freely from a warm to a cool environment. The second part is called entropy. Entropy is the amount of disorder in a system; as a system gets closer and closer to equilibrium, it also becomes more disordered, increasing its entropy. It is the intertwining of both parts that is the second law of thermodynamics. The following example is an analogy that uses something as simple as a neatly stacked deck of cards (symbolizing the energy of an object) dropping to the ground (representing entropy).

Condensation is formed when warm water vapor collides with the cooler molecules in the glass (see next page).

When stacked neatly, the cards are ordered and have very little entropy (they are warm). However, when allowed to fall to the floor (as heat naturally flows to a cooler environment), the cards are disordered and have more entropy than before (they are now cooler and do not have as much energy as when they were stacked). The only way to give them order again is to pick each card up (do work, thereby increasing temperature) and give them back the order they once had.

The process of condensation and evaporation all employ the second law of thermodynamics. The following questions address these processes.

On a warm day, why do **water droplets** accumulate on the outside of glasses and soda bottles?

The water does not seep through the container, but instead comes from the air surrounding it. When warm water vapor, which has a great deal of internal energy, collides with the slower and cooler molecules in the container, the container absorbs a great deal of the vapor's energy. This cools and slows down the water vapor, changing it from a gas to liquid water droplets.

What is another example of **condensation**?

Condensation can be found on the insides of windows in warm, humid homes. The high-energy water vapor, when colliding against the cool, usually single-pane windows, slows down and changes into water droplets.

How do liquids **evaporate**?

Molecules do not have to boil into a gas to leave the liquid state. Instead, energetic molecules can leave through a process called evaporation. All liquids have a "surface tension"; a rough analogy would be that all liquids have a denser, tough rind covering them, like the skin of a fruit. In order to evaporate from the liquid, a molecule must penetrate this tough skin, and doing so takes energy. So if a molecule has more kinetic energy than the amount of energy it takes to penetrate the "skin," it can pass through the skin, and will lose energy in doing so. This means that only the molecules with high velocities will be able to evaporate. Higher velocities mean higher temperatures, so by removing the molecules with the highest temperatures from a liquid, the temperature of the remaining liquid is lowered.

How can evaporation be a **cooling** process?

If not for evaporation, our bodies would overheat quickly. Whenever we exert ourselves, our bodies perspire in order to place a liquid on our skin that will evaporate. The evaporation process quickens when it occurs in

How do refrigerators keep food cool?

Most food items need to stay cool to slow down the growth of harmful bacteria that can cause spoilage. Milk and other dairy products, for example, would spoil in just a few hours if it were not for refrigeration. The way a refrigerator keeps items cool is by removing warm air from inside the refrigerator through the process of evaporation.

In the back of most refrigerators lies a coil of heat-exchanging pipes. A compressor pressurizes Freon or some other coolant, and pushes the gaseous coolant into the pipes at extremely high temperatures. Upon entering these pipes, the gas quickly cools and condenses into its liquid form by releasing heat into the room as it travels through the pipes. To do this, the cool liquid absorbs heat from the food and air inside the refrigerator. After obtaining such huge amounts of heat, the liquid Freon actually begins to boil inside the tubes. The compressor then pressurizes the Freon and the process begins all over again.

a warm environment, such as on our skin. By absorbing the heat from our skin, the molecules in the sweat will have enough kinetic energy to break free from our bodies and evaporate into the air. The perspiration's absorption of heat from our skin and quick evaporation provide our bodies with an effective cooling system.

What is **Freon,** and why is it no longer used in new refrigerators and air conditioners?

Freon is a refrigerant that was first developed by the DuPont Company of Wilmington, Delaware. Freon was considered a major breakthrough for refrigeration for it was non-reactive, meaning it was safe to have in the kitchen. Prior to the development of Freon, dangerous chemicals such as ammonia and ether were used as refrigerants. A type of chlorofluorocarbon (CFC), Freon carries chlorine atoms to the upper atmosphere, which contributes in the destruction of the ozone layer, an essential barrier against harmful ultraviolet sunlight. Freon's destructive nature has been known since the 1970s, but it was not until the early '90s that legislation was implemented banning the use of Freon in new air conditioners and refrigerators. Unfortunately, when the chlorine destroys an ozone molecule, the chlorine is not destroyed, but instead continues to live for a while destroying more ozone. In fact, more Freon is still headed toward the upper limits of the atmosphere, for it can take several years for Freon to reach such elevations.

What is being used **instead of Freon**?

In order to be an effective cooling agent in a refrigerator or air conditioner, the liquid refrigerant must easily evaporate. Chlorofluorocarbons were the ideal fluid, replacing the toxic ammonia that was used before DuPont developed Freon. Today, safer fluids known as hydrofluorocarbons have been developed, eliminating the use of chlorine in cooling fluids. However, although the hydrofluorocarbons are safer for the environment, they are not as energy efficient as the CFCs were.

What is a **Carnot cycle**?

A French physicist named Nicolas Leonard Carnot said that an ideal engine would convert all the heat into usable mechanical energy; he also stated that such an engine would be impossible to construct for the engine would need to be reversible, so that all the energy converted during the engine's cycle could be returned to its original state. Engines have been made that are extremely close to the ideal "Carnot efficiency," lacking only the losses due to friction of moving parts.

THE THIRD LAW

What is the **third law** of thermodynamics?

The third law of thermodynamics states that absolute zero, the lowest possible temperature, the point at which there is no energy, can never be reached. Scientists have seen this through experimental results in the laboratory. As mentioned earlier, physicists have been able to achieve temperatures as low as millionths of Kelvin, but have never been able, and according to the third law of thermodynamics, will never be able, to reach absolute zero.

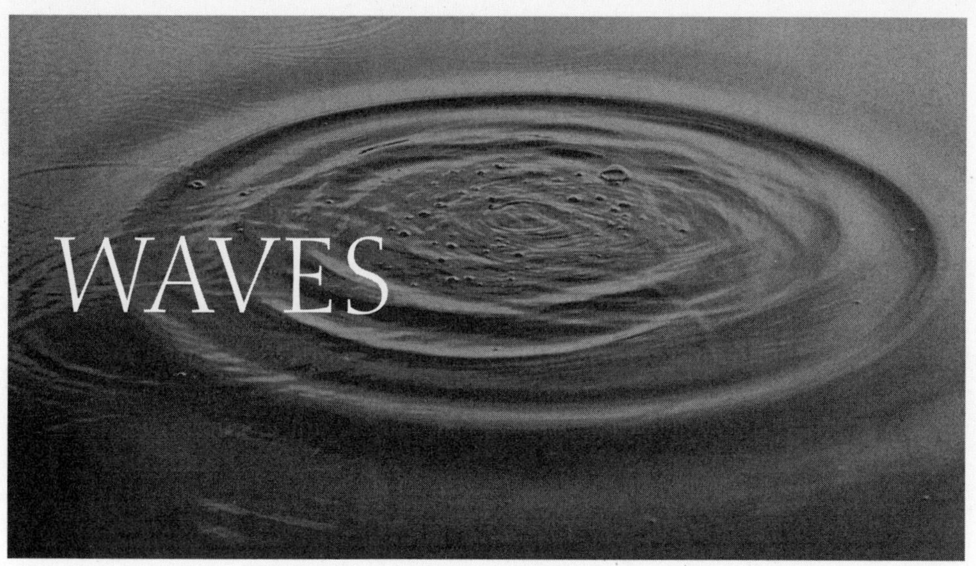

WAVE PROPERTIES

What is a **wave**?

A wave is a traveling disturbance that moves energy from one location to another without transferring matter. Oscillations in a medium or material create mechanical waves that propagate away from the location of the oscillation. For example, a pebble dropped into a pool of water creates vertical oscillations in the water, while the wave propagates outward horizontally along the plane of the pond.

What are the two major **classifications** for waves?

Transverse and longitudinal waves are the two major classifications for waves in physics. A transverse wave can be created by shaking a string or rope up and down. Although the string is moved up and down, the energy from the oscillations moves perpendicular and away from the vibrating source.

The oscillations in longitudinal waves do not vibrate perpendicular to the direction the wave is traveling, but instead oscillate in the same direction the wave is moving. The medium in longitudinal waves pushes close together at some points (compression) and separates from each other immediately after (rarefaction). The best example of longitudinal waves are sound waves, which are a series of back and forth, longitudinal oscillations of air molecules that compress and rarefact in a medium such as air or water.

What determines the **velocity** of a wave?

The velocity of a wave depends upon the material or medium in which it is traveling. When a wave enters a new medium, the elasticity and density of the medium may cause a change in the velocity of that wave. Typically, the denser and more elastic the medium, the faster the wave will travel.

Once the wave is in the particular medium, all waves of that type will travel at the same speed. For example, a sound wave traveling through air at 0°C will travel at 331 m/s (meters per second). Regardless of the sound's frequency, the sound will always travel at 331 m/s until a change in the medium is encountered.

What are some of the **terms used to define the properties of waves**?

Type of Wave	Term	Definition
Transverse	crest	The highest point of the wave.
	trough	The lowest point of the wave.
Longitudinal	compression	An area where the material or medium is condensed as a result of a force applied to the medium.
	rarefaction	An area that follows a compression where the material or medium is spread out.
Transverse & Longitudinal	amplitude	The distance from the midpoint to the point of maximum displacement (crest or compression).
	frequency	The number of vibrations that occur in one second; the reciprocal of the period.
	period	The time it takes for a wave to complete one full vibration; the inverse of the frequency.
	wavelength	The distance from one point on the wave to the next identical point; the length of the wave.

What is the relationship between **frequency, wavelength, and velocity**?

As long as a wave remains in one medium, the speed of a wave will remain constant. Since the velocity of a wave does not change, under those conditions, the only variables that could change would be the frequency and the wavelength. The equation for wave speed is: velocity = frequency × wavelength. Therefore, if the frequency of a wave increased, the wavelength would have to decrease in order for the velocity to remain constant. The frequency and wavelength are inversely proportional to each other.

For example, sound travels at a speed of 331 m/s (meters per second) in air that is at the freezing mark. If the frequency of different sounds waves were changed, the wavelengths would change as follows:

Velocity of Sound (0°C)	Frequency (Hz)	Wavelength (m)
331	128	2.59
331	256	1.29
331	512	0.65
331	768	0.43

What is the relationship between **frequency and period**?

Frequency is how many cycles of a vibration occur per second and is measured in cycles per second or hertz (Hz). The period of a wave is the amount of time it takes a wave to vibrate one full cycle. These two terms are inversely proportional to each other.

For example, if a wave takes one second to vibrate up and down, the period of the wave is 1 second. The frequency is the reciprocal of that, and is 1 cycle/sec because only one cycle occurred in a second. If, however, a wave took half a second to vibrate up and down, the period of that wave would be .5 seconds, and the frequency would be the reciprocal of that wave, resulting in a frequency of 2 cycles per second. As a result, it is a good idea to keep in mind that a wave with a long period has a low frequency, while a wave with a very short period has a high frequency.

OCEAN WAVES

See also: MOVEMENT chapter, "TIDAL FORCES"

What **type of wave** is an ocean wave?

Ocean or water waves look like transverse waves, yet are actually a combination of both transverse and longitudinal waves. The water molecules in a water wave vibrate up and down in tiny circular paths. The circular path of the water wave creates an undulating appearance in the wave. At the crest of the wave, the water molecules tend to spread out a bit, resulting in an area of rarefaction, while in the trough of the wave, the water molecules are compressed.

How fast does the **wind** need to **blow** in order to produce different types of waves?

Wind rubbing against the water surface is a major cause of waves. Since the water cannot keep up with the wind velocity, the water rises and then falls, creating the familiar wave-like motion. Depending upon the wind velocity and the distance the wind has been able to travel over the water, different size waves are generated.

Type of Wave	Wind Velocity	Effect
Capillary waves	Less than 3 knots	Tiny ripples. The longer they vibrate, the larger they can become.
Chop or regular waves	3–12 knots	Combined capillary waves that have traveled far and formed larger waves.
Whitecaps	11–15 knots	Amplitude of wave must be over 1/7th the wavelength in order to break into a whitecap.
Ocean swells	No specific speed	Form over long distances from a compilation of different waves joining together.

How is the speed of an ocean wave determined?

The speed of an ocean wave is determined by the distance between two successive crests or its wavelength. The longer the wavelength, the faster the wave travels. A small surface wave, such as a ripple created by the wind, travels quite slowly because it has such a short wavelength. A tidal wave, in contrast to a ripple wave, is created by a seismic disturbance in the ocean floor, and has a very long wavelength and can travel at extremely high velocities. The velocity of the wave is also directly proportional to the amount of energy that the wave carries, which explains why a tidal wave can cause so much damage to a shoreline.

Why do ocean waves **break** as they **approach the beach**?

Ocean waves rarely break before they come in contact with a cliff or mountainside shoreline. Ocean waves only seem to break as they approach a gradual decrease in depths, such as beaches. A shoreline with a gradual decrease in depth will produce a more spectacular break than a wave that encounters a steep decrease in depth.

The reason waves break has to do with the velocity of the wave and the depth of the water. An ocean wave with a large velocity has a long wavelength and large amplitude. As the wave moves toward the beach, it wants to continue traveling at a constant velocity. Unfortunately for the wave, as the ocean depth begins to decrease, the bottom of the wave gradually encounters more and more friction, causing the lower section of the wave to travel slower than the upper section of the wave. As the lower section decreases its velocity, the inertia of the crest carries it over the trough. When there is not enough water underneath it to support the crest, the wave breaks.

How is **surfing** a lot like downhill skiing?

Surfing is different from skiing in that surfing is usually a bit warmer than skiing. However, there is one major similarity between the two

Surfer under a tubular wave.

sports: the fact that both sports require the athlete and board to travel down a hill. In skiing, the hill is a mountain covered with snow, while in surfing, the hill is the rising water of a breaking ocean wave. An ideal surfing wave has a great deal of energy that encounters an extremely gradual decrease in ocean depth as it approaches the beach. While the surfer moves down the wave, the water on the front edge of a crest continually rises underneath the surfer, allowing the surfer to ride down the wave without actually moving downward. The surf continues until the wave losses its energy and finishes breaking.

Where are the **best surfing beaches**?

The best surfing beaches are located in oceans where the wavelengths are quite large. The wavelength helps determine the speed of the waves. The longer the wavelength, the faster the wave. Another determinant of a good surfing beach is if the depth of the shoreline is gradual or steep. Waves tend to break when the depth of the water is 1.3 times the height of the wave. Therefore a beach with a long, gradual decreasing water depth is the best for good surfing.

Some of the best surfing is done out in Waikiki in Hawaii and on the west coast of the U.S. The Pacific Ocean, famous for its long wavelengths and gradual decreasing depth beaches, has some of the best surfing in the world.

What is a **tidal wave**?

A tidal wave or tsunami is not caused by windy conditions or tides, but instead by underwater earthquakes and volcanic eruptions. The seismic disturbances create huge upward forces on the water, the opposite of dropping rocks into water. Large but rare tsunamis can be quite destructive upon reaching the shoreline due to their large wavelengths, but most seem to be only a meter or two high.

One of the most famous tidal waves occurred over a century ago in Indonesia. A tsunami, traveling at a speed of several hundred kilometers per hour, crashed into the shoreline, killing thirty six-thousand inhabitants. Hawaii has also been a favorite target of tsunamis. Over the past two hundred years as many as forty tsunamis have struck the Hawaiian Islands.

ELECTROMAGNETIC WAVES

What is an **electromagnetic wave**?

Light, radio, and X-rays are all examples of electromagnetic waves. Electromagnetic waves are a special type of transverse wave; they consist of two perpendicular transverse waves, one component of the wave being a vibrating electric field while the other is a corresponding magnetic field.

Although all electromagnetic waves travel at the speed of light, the particular type of electromagnetic wave can be defined by its frequency or wavelength. Finally, electromagnetic waves differ from other transverse and longitudinal waves in that they do not need a medium such as air, water, or steel to travel through. Radio, gamma, and visible light waves have no problem traveling through the emptiness of space.

An electromagnetic wave.

How is an electromagnetic wave **created**?

Electromagnetic waves are created by moving charges in atoms, which create an electric field that in turn creates a corresponding magnetic field. The energy from the moving electrons radiates (not necessarily uniformly) into the area around the moving charge.

What is the **electromagnetic spectrum**?

The electromagnetic spectrum lists the wide range of electromagnetic (EM) waves from low to high frequency. The spectrum ranges from radio waves, which have the lowest frequency in the spectrum, all the way to gamma rays, which have a very high frequency. In the middle of the spectrum is a small section that comprises the frequencies for visible light.

Who predicted electromagnetic waves?

In 1861, James Clerk Maxwell studied and demonstrated the mathematical relationship between vibrating electric and magnetic fields. In his

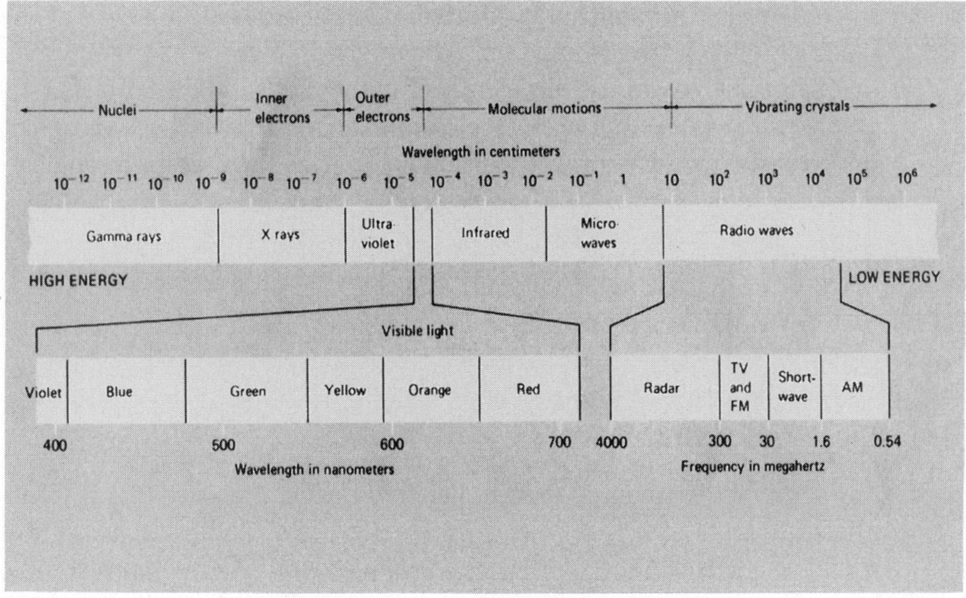

The spectrum of electromagnetic radiation.

Treatise on Electricity and Magnetism, written in 1873, Maxwell described the nature of electromagnetic fields and waves through four differential equations, known to physicists today as "Maxwell's Equations." Although Maxwell never proved his theory in the laboratory, he is credited with predicting the existence of the special wave.

Maxwell was a professor at Cambridge University in England from 1871 until his death in 1879. He published other works on thermodynamics and the motion of matter as well. He also developed the kinetic theory of gases, and performed a great deal of research in the field of color vision. Although Maxwell is not widely known to the lay audience, he is revered in the scientific community, and rates in the pantheon of physics greats with Newton and Einstein.

Who proved that the electromagnetic wave actually existed?

It wasn't until Heinrich Hertz designed a radio transmitter and receiver that the electromagnetic wave was truly discovered to exist beyond pure mathematical theory. Through his research, Hertz proved that electrical signals can be transmitted by electromagnetic waves that travel at the speed of light. It was Hertz's breakthroughs in electromagnetic waves that paved the way for the development of the radio and the wireless telegraph. As a result of his work on electromagnetic waves, Hertz was given the honor of having his name, hertz (Hz), used as the unit for frequency.

RADIO WAVES

Are radio waves **sound waves**?

Although radios are often used to listen to music, the wave that is transmitted to the radio is an electromagnetic wave. Radio waves are not sound waves, yet in some instances, they do carry information to a radio to produce sound waves. Once the antenna receives the radio wave, the circuitry inside the radio changes the electromagnetic wave into an electrical signal, which is sent to speakers in order to convert the electric signal into the sound waves that our ears receive.

How do **antennae** transmit and receive radio signals?

Antennae for radio and television signals are used to either transmit or receive electromagnetic radio waves. Transmitting antennae cause the electrons to vibrate; this oscillating electric field creates an oscillating magnetic field, resulting in the propagation of electromagnetic waves. When the receiver is tuned to a particular frequency, the radio wave induces an electric current in the receiving antenna, which is sent to the radio receiver.

Who invented the **radio**?

In 1895 Guglielmo Marconi, a twenty-year-old Italian inventor, created a device that transmitted and received electromagnetic radio waves over a one-kilometer distance. Later improvements to his antenna and the development of a crude amplifier enabled him to receive a British patent for his wireless telegraph. In 1897, he transmitted signals to ships 29

Does the dimension of an antenna play a significant role in the reception of a radio wave?

The length of an antenna determines the frequency that it best receives. The general rule for radio and TV antennae is that the length of the antenna should be half the wavelength of the wave it is attempting to receive. This allows the induced electrical current in the receiving antenna to resonate at that particular frequency.

The exception to this rule is the loop or coil antenna. Magnetic wire loop antennae are found inside transistor radios and receive only low-frequency radio waves found on the AM band. A half-wavelength straight wire antenna would have to be extremely long in order to receive the low-frequency AM band radio waves. Instead, the coiled antenna inside a transistor radio reacts to the vibrating magnetic field of a radio wave, which in turn induces a large current.

Home radio and television antennae typically have a broad bandwidth but little gain. A broad bandwidth allows the antenna to receive a larger array of frequencies than otherwise would be possible with a narrower bandwidth. However, a broader bandwidth sacrifices the gain or sensitivity of the antenna to only narrow-frequency bands.

kilometers from shore and just four years later was able to send wireless messages across the Atlantic Ocean. As a result of Marconi's work on radio transmitters and receivers, Marconi was the co-winner of the 1909 Nobel prize in physics.

What are **Mhz, kHz, and Ghz** after the frequency numbers?

The unit for frequency is hertz (Hz), named after the German physicist Heinrich Hertz, the discoverer of electromagnetic waves. Hertz represents the number of vibrations or cycles per second of a wave. On radio

Guglielmo Marconi with his wireless system (see page 182).

receivers, the letters kHz, Mhz, and Ghz are often found after the frequency number. KHz represents kilo or one thousand hertz; Mhz, or megahertz, is one million hertz; and Ghz is gigahertz, or one billion cycles per second.

What are the different frequencies of radio waves and microwaves that allow for **communication**?

The following table outlines the different sections of the radio and microwave spectrum and their uses:

Frequency Range	Name & Abbreviation	Use
3–300 Hz	Extremely low frequency (ELF)	Telegraph, teletypewriter
300–3 kHz	Voice frequency (VF)	Telephone circuitry
3 kHz–30 kHz	Very low frequency (VLF)	High fidelity
30 kHz–300 kHz	Low frequency (LF)	Maritime mobile, navigational radio broadcasts
300 kHz–3 Mhz	Medium frequency (MF)	Land & maritime mobile radio, radio broadcasts
3 Mhz–30 Mhz	High frequency (HF)	Maritime & aeronautical mobile, amateur radio
30 Mhz–300 Mhz	Very high frequency (VHF)	Maritime & aeronautical mobile, amateur radio, television broadcasts, meteorological communication
300 Mhz–3 Ghz	Ultrahigh frequency (UHF)	Television, military, long-range radar
3 Ghz–30 Ghz	Superhigh frequency (SHF)	Space and satellite communication, microwave communication
30 Ghz–300 Ghz	Extremely high frequency (EHF)	Radio astronomy, radar

AM AND FM

What does **AM** mean?

AM, or amplitude modulation, is a method of transmitting information with a radio wave. Although the radio wave does not transmit sound waves, it carries the information that is needed to produce a particular sound wave. Sound waves are longitudinal waves and are created by compressions and rarefactions in air. An amplitude modulated signal represents the amount of compression and rarefactions by changing or modulating the amplitude of the radio wave that is emitted. To represent a compression, a high-amplitude radio wave is created. To represent a rarefaction, a low-amplitude radio wave is emitted. The radio receiver measures the difference in the amplitude and sends the information to the speakers to emit the appropriate sound wave.

Why do FM stations usually sound better than AM stations?

FM stations produce a better sound than AM stations because FM stations transmit the radio waves at full power. Amplitude modulation changes the amplitude of the radio wave to inform the receiver of how much compression and rarefaction is needed from the speakers. For this reason, the AM signal never transmits at full power. FM, on the other hand, keeps its signal at full power and instead varies the frequency slightly. The lower power level from AM signals allows the receiver to pick up other electromagnetic waves, because the receiver cannot differentiate between the real signal and the extra electromagnetic noise. The FM band receiver can differentiate between the extra noise and the signal because the transmitted signal overpowers the extra background noise.

What does **FM** mean?

Frequency modulated or FM radio waves represent the compressions and rarefactions that the speaker must emit through slight variations in the frequency of the radio wave. In order to represent a compression, the frequency of the wave is increased slightly, while to represent a rarefaction from the speaker, the frequency of the wave is slightly decreased. Individual FM stations have broader ranges than AM stations to accommodate the slight fluctuations in the frequency without interfering with a neighboring station's frequency.

Where is AM and FM found in the **electromagnetic spectrum**?

Amplitude modulated (AM) radio is located between frequencies of 550 kHz and 1600 kHz. Frequency modulated (FM) radio is located between the frequencies of 88 Mhz and 108 Mhz. Other radio frequencies such as police bands, TV bands, and shortwave communications use AM and FM methods of communication to transmit information as well.

What **other systems** besides FM band radio **use frequency modulation** to transmit information for sound?

Besides the frequencies between 88 Mhz and 108 Mhz on the FM band, other broadcasting frequencies use frequency modulation for transmitting information at full power, as opposed to the varying power in amplitude modulation. Sound for television, mobile cellular telephone systems, and microwave radio systems use the frequency modulation to transmit information for high-fidelity sound. Since these frequencies are on the high end of the radio spectrum, they only have a line-of-sight range to use the FM effectively.

Why are many **microwave transmission systems changing** from FM to AM?

Some high-frequency microwave transmissions have changed from FM to AM transmissions because the range of fluctuation at such a high frequency is so small that interference with other channels is just as prevalent as it is with AM. For this reason, along with technological improvements in AM broadcasting, many high-frequency microwaves transmissions have changed to what is called single-sideband or SSB/AM transmissions. The SSB/AM allows the microwave system to transmit more than three times the number of voice signals that a conventional FM microwave system can. However, with the constant improvements in technology, this system is being changed over to Pulse Code Modulation, a form of digital communication that should allow an even greater number of instantaneous signals to be transmitted.

How does an **FM** band station transmit **stereo sound**?

Stereo sound means that two separate sounds emit from a pair of speakers. A radio wave only transmits one frequency at a time, making it seem very difficult to produce different types of sounds from two speakers.

According to the Federal Communications Commission (FCC), FM frequencies can only produce sound waves in speakers between 50 Hz and 15,000 Hz (the range for human hearing is 20 Hz to 20,000 Hz). Although radio signals above 15,000 Hz cannot be produced as a sound from the speaker, the receiver may still receive such high-frequency messages. Stations wishing to broadcast in stereo transmit a "pilot signal" to the receiver at 19,000 Hz, which carries the information that the receiver needs to broadcast in stereo.

What are the **ranges** for FM and AM bands?

All electromagnetic waves travel in a straight line while in a uniform medium such as the lower atmosphere. Therefore, most radio waves only have what is called a line-of-sight range. That means that if a mountain range or the curvature of the Earth were in the way of the radio signal, the receiver would be out of range and would not receive the signal. This is why most broadcasting antennae are placed on tall buildings or mountains to help increase the line-of-sight range.

However, lower-frequency radio waves (those with frequencies below 30 Mhz) are able to reflect off the charged particles in the Earth's ionosphere; this is referred to as "skip." Instead of passing through the ionosphere and entering space as higher-frequency electromagnetic waves do, the lower frequencies on the AM band have the ability to reflect back toward the Earth and increase their range dramatically. As an added bonus, the reflective conditions of the ionosphere in the evening hours are superb and can increase range up to thousands of kilometers from a transmitting tower. Although FM stations tend to be of higher fidelity, AM broadcasts tend to have a much larger range.

MICROWAVES

How are microwaves used for **communication**?

Microwaves are superhigh frequency (SHF) electromagnetic waves in the frequency range of 3 Ghz to 30 Ghz, and are often used to transmit tens of thousands of signals across a very narrow band of frequencies. Microwaves, which are usually used to send telephone, television, radar, and meteorological communications, are created in klystron and magnetron electron tubes. Microwaves are known for reliability and error-free

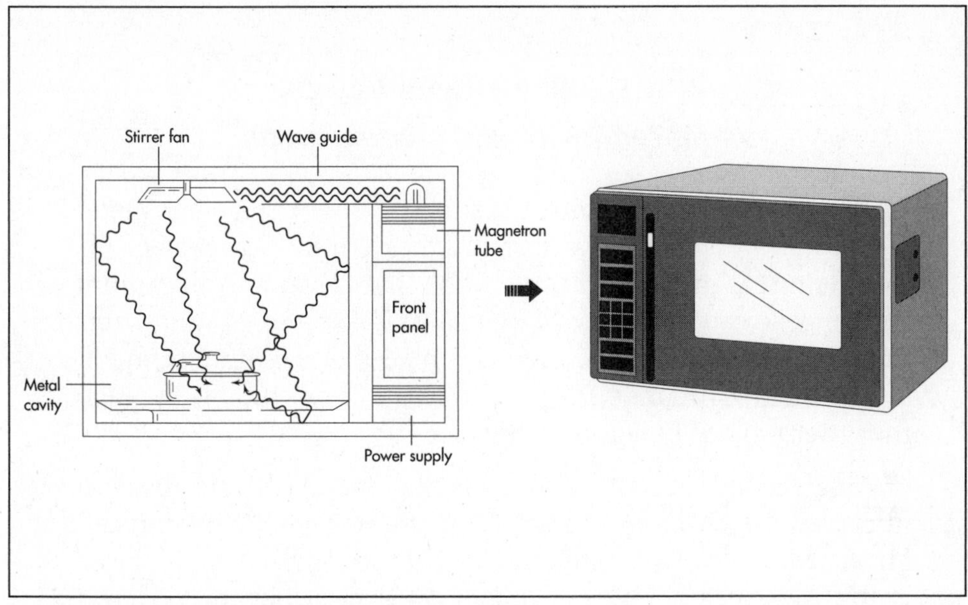

In a microwave oven, a magnetron tube creates microwaves, which the stirrer fan directs toward the oven cavity. The waves bounce off the metal walls until they are absorbed by the food, heating it.

transmissions, but they are difficult to send long distances due to interference from mountains, buildings, and the curvature of the Earth itself.

Microwaves can be transmitted one of two ways. The first is the line-of-sight approach, where the microwave transmitter is pointed at a microwave receiver (these must be no more than 30 km apart). The second method of transmission is to send the signal up to a satellite that reflects it back down to a receiving dish.

What are **other uses for microwaves** besides communication?

In addition to having a great range of frequencies to transmit information, microwaves are used every day in kitchens around the world. Microwave ovens generate superhigh frequency waves and scatter them throughout the oven. The frequency of microwaves excite water molecules into resonance and cause them to collide with one another. Friction generated by the collisions changes the kinetic energy of the water into heat that warms the food. Anything consisting of water can be heated by a microwave oven.

Why shouldn't metal objects be placed inside microwave ovens?

Manufacturers caution consumers about placing metal containers and aluminum foil inside microwave ovens for two main reasons. The first reason is that metal and aluminum may impede cooking. Microwaves warm food by resonating water molecules within the food. If food is placed under aluminum foil or in a metal container, the microwaves will not be able to reach the water molecules and cook the food.

The second reason for not placing metal objects inside a microwave oven is for the safety of the microwave oven itself. Metal acts as a mirror to microwaves. If too much metal is placed in the oven, the microwaves will not get absorbed by the food, and will bounce around until they overload the oven and cause damage to the magnetron, the device that produces the microwaves.

What is the function of the **grating on the door** of a microwave oven?

People using microwave ovens want to see the food cooking inside the oven, yet not be bombarded by potentially harmful microwaves. In order to prevent the escape of microwaves through the plastic or glass door, a grating consisting of small holes is used to reflect the microwaves back into the oven. The microwaves (which have a wavelength of about 12 centimeters) are too big to pass through the holes, but visible light, whose wavelength is smaller than the opening, can easily pass though the grating. Although the grating protects people from the microwaves, some microwaves can still leak out through the door seal, however, if it's not cleaned occasionally.

Can a microwave oven be used to **dry things**?

Since water molecules are warmed and eventually boil off by microwaves, anything that is wet can be dried in the microwave. How-

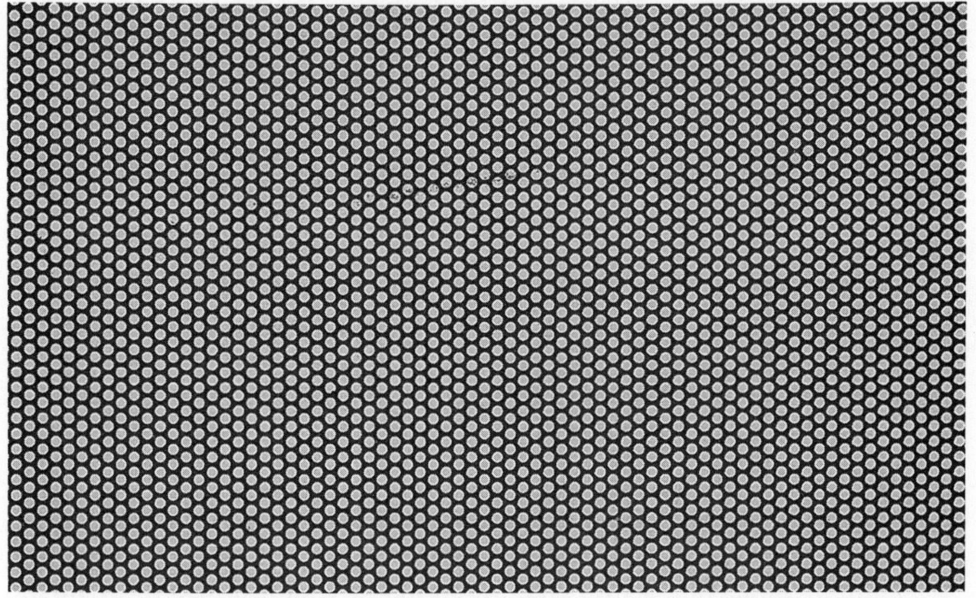

A close-up view of the protective grid on a microwave oven, which keeps the waves from escaping because the holes in the grid are smaller than the wavelength of the microwaves.

ever, there is one very important consideration that must be made before placing the object inside a microwave—the object being dried must not contain a great deal of water itself. Microwaves are wonderful at drying wet books, papers, and magazines, but must never be used to dry things like plants or small animals. Living things would be killed by the resonation of water molecules *inside* their bodies.

SHORTWAVE RADIO

What is **shortwave radio communication**?

Shortwave or "ham" radio are ranges of frequencies set aside for amateur radio operators around the world. Shortwave radio frequencies are mostly in the high-frequency range, but some other "ham" frequencies are located in the medium frequency and ultrahigh frequency ranges as well.

INTERFERENCE, SUPERPOSITIONING, AND RESONANCE

Does a water, sound, or mechanical wave **travel forever**?

If a wave were traveling in a frictionless environment, such as an electromagnetic wave traveling through the vacuum of space, the wave would travel at a constant speed until it encountered a medium. However, on Earth there is a great deal of friction. Friction causes a gradual decrease in the amplitude of a wave by changing some of its energy to thermal energy. This gradual decrease in a wave's amplitude is known as damping.

An example of damping in waves can be easily observed with sound waves. Over time and distance the frictional effects from the surrounding air molecules cause the amplitude of sound waves to gradually decrease.

Electric waves also experience damping. Electric waves traveling as transverse waves through wires will eventually decrease in energy due to the frictional effects of imperfections in the wires. Over distance, the amplitude of an electric wave might decrease so much that the receiver of the wave would not be able to register the electrical impulses. In order to correct such problems, the amplitude of electric waves must be increased so the signals will not be lost. Amplifiers are typically used to increase the amplitude of the wave before damping reduces the wave to nothing.

SUPERPOSITIONING

What is the difference between **constructive and destructive interference**?

Interfering waves do not crash and destroy one another; instead, they interact and pass through each other. The interaction between waves is called superpositioning, and can be predicted through the addition or subtraction of amplitudes when they interact with each other.

What are dead spots in auditoriums?

Poorly designed auditoriums can have dead spots. Dead spots are places where destructive interference occurs from the interaction of two or more sound waves. For example, a soloist on stage sends sound waves into the audience. Some of the waves hit the walls of the auditorium, while other waves travel directly to the listeners. In some situations, a wave can interfere with a reflected wave so perfectly as to destructively interfere and cancel each other out at that particular location. As a result, the listeners seated in those particular seats would hear nothing from that soloist. Someone sitting a few seats over from the dead spot, however, might not experience the destructive interference and would hear the soloist just fine. In addition, the person seated in the dead spot for the trumpeter might be able to hear the trombone player because the waves from the trombone are not traveling from the same source of sound and thus are not causing destructive interference. (Refer to the chapter on SOUND for handy answers dealing with acoustical engineering.)

A very large positive amplitude or constructive interference occurs if two positive amplitudes interfere or a large negative amplitude is created if two negative amplitudes interfere. A flat-line or destructive interference results if two opposite amplitudes interact with one another. After the waves interfere with each other, the individual waves will continue to travel with the same velocity that they had before interfering with each other.

What is a **rogue**?

A rogue is a pair of water waves, created by the same storm, that meets and forms constructively interfering waves. These waves travel in different directions because of quick changes in the pathway of the storm; when they meet, they form a huge positive amplitude or a huge negative

The B-52 Stealth Bomber.

amplitude. In a British sailing regatta in 1979, fifteen people died when a fifty-foot rogue destroyed dozens of sailing vessels.

How are today's fighter jets using destructive interference to **mislead enemy radar**?

The new French fighter plane known as Rafale uses a device to help the jet evade radar. Radar (an acronym for "radio detection and ranging") is an effective navigation and early warning system for military forces. It can detect foreign objects by sending electromagnetic waves out into the atmosphere and measuring the time and frequency of the reflecting wave after it has struck that object. In order to evade radar, the Rafale is using technology called active cancellation, which receives an incoming wave and sends out the direct opposite pattern of that wave, in this case a radar wave half a wavelength out of phase with the incoming radar. When the two waves interfere with one another, the waves experience destructive interference canceling out the signal. Because there is no return signal, the enemy can't figure out where the plane is located.

What is a **stealth plane**?

Stealth aircraft are planes that are able to avoid radar detection. Their peculiar shapes and angles deflect radar waves away from the plane, or in some instances, the plane's outer fuselage can actually absorb the radar waves without reflecting it back to the enemy radar transmitter. (Refer to the FLUIDS chapter for more information about aerodynamics and aviation.)

What is a **standing wave**?

A standing wave occurs when a continuous set of waves reflect off a surface and overlap each other. If the frequency is set just right for a perfect overlap of the initial and reflected waves, the wave will seem to stand still. Two distinct sections on a standing wave become evident. Sections that stay still are called nodes and sections between nodes that move up and down dramatically are called anti-nodes. To create a standing wave, the frequency of the waves and the distance the waves travel before reflecting must be adjusted just right in order to make the wave appear to "stand."

How are **nodes and anti-nodes** created in a standing wave?

Initial waves overlapping reflected waves will create two sections of a standing wave. The first section occurs where constructive interference takes place. The constructive interference is a result of crests overlapping crests or troughs overlapping troughs, resulting in a large changeover between a large positive amplitude and a large negative amplitude. This section of constructive interference is called an anti-node or peak. The other section of interference in a standing wave is formed by destructive interference. Points of destructive interference, called nodes, are created by crests and troughs continuously interfering and canceling each other out. Nodes do not move up and down since the initial and reflected waves produce a flat-line effect.

How are standing sound waves generated in **musical instruments**?

Many instruments depend on standing waves to produce their sound. Standing waves are created in the vibrating air inside a pipe organ, on the strings of a guitar or violin, and in the air columns of a trumpet or flute. In order to change the tune of an instrument, the standing wave inside the instrument must be altered. By changing the length of a wind instrument, or the tension and length of the strings for a string instrument, a different frequency standing wave is produced, which creates the different musical notes.

What is **natural frequency**?

All objects that posses some elasticity have a natural frequency. Natural frequency is achieved when the smallest amount of energy is needed to continue a vibration in an object. The natural frequency of the object depends chiefly upon its physical characteristics and especially the elasticity of the object itself.

RESONANCE

What is **resonance** and how can it be achieved?

Resonance occurs when the frequency of a continuous wave achieves a standing wave with maximum amplitude. To achieve resonance, a force must continuously vibrate an object at the natural frequency of that object. At this point, resonance occurs, and then very little force is needed to keep the resonance alive.

Where can resonance be found on the **playground**?

Children discover resonance early in life. When playing on swings, they use their arms and legs to pump themselves back and forth on the swing. When the child reaches a particular frequency on the swing, they discover that they do not have to pump as hard as before. At this point, they have matched the natural frequency of the swing, because they need only to put in very little force to keep them swinging with maximum amplitude. However, the resonance achieved on the swing can be destroyed if the child pumps or a parent pushes at the wrong time. As long as the force provided to the swing matches its natural frequency, maximum amplitude and resonance are maintained.

How can resonance cause **crystal glasses to break**?

Ella Fitzgerald performed this physics experiment in an advertisement for Memorex audio cassettes. The ad claimed that the famous singer could create a pure tone at just the right frequency to cause a crystal wine glass to break. The frequency she produced was the natural frequency of the glass.

When the amplified sound waves from Ella Fitzgerald pushed on the molecules in the glass, some of the sound's energy was transferred from the sound's kinetic energy into the glass' kinetic energy. The molecules inside the glass vibrated more and more until resonance was achieved. When the resonant frequency was achieved, it caused a huge amplitude inside the goblet resulting in the shattering of the glass.

How did resonance destroy the **Tacoma Narrows Bridge** in Washington State?

The Tacoma Narrows Bridge, or "Galloping Gertie" as it was often called, was built in 1940 and was known for its unusual, undulating movement. All bridges vibrate to some extent, but to many motorists, the suspension bridge in Tacoma felt more like an amusement park ride than a bridge.

On the morning of November 7, 1940, four months after the bridge opened, the wind was blowing at approximately 42 miles per hour. This moderate wind hit the solid girders of the bridge deck and caused the deck to vibrate back and forth as it did almost every day since the bridge had opened. However, to the shock of engineers and spectators alike, the bridge began to vibrate more dramatically than ever before. It appeared as though a standing wave had formed between the two towers of the bridge. There was one distinct node in the center of the bridge and an anti-node on each side of the node.

Throughout the morning, the amplitude of the torsional standing wave grew, indicating that the bridge was achieving resonance. After several

The Tacoma Narrows Bridge, which collapsed on November 7, 1940, after it began vibrating at its natural frequency, leading to resonance, which caused its collapse.

hours of dramatic vibrations, the bridge deck collapsed into the river below, along with its only casualty, a dog named "Tubby," left in a car by its owner, who narrowly escaped death himself.

Engineers feel the collapse was not the result of high winds, but instead the fact that the wind caused the bridge deck to vibrate at its natural frequency. The natural frequency was determined not only by the material with which the bridge was constructed, but by the distance between the towers of the bridge as well, which was equal to one full wavelength of the oscillations. When an object vibrates at its natural frequency for a long enough period of time, resonance can occur, in this case causing the bridge to collapse. Today, civil engineers study the Tacoma Narrows Bridge collapse to make sure that nothing similar ever happens again. (Refer to the OBJECTS AT REST chapter for more information on bridges).

What is a **torsional wave**?

A torsional wave, such as the Tacoma Narrows Bridge created, is a wave that is not only displaced vertically, but twists in a wave-like fashion as well. The torsional wave on the Tacoma Narrows Bridge achieved resonance in two orientations. The first resonance was seen as the undulating movement that took place over the length of the bridge, while the second resonance, seen as the twisting motion, occurred from side to side on the bridge.

How do **organ pipes** create sound?

An organ pipe produces sound by resonating the air molecules inside the pipe. The air is vibrated by blowing across a slit in the side of the pipe, resulting in pressure differences that cause the rest of the air molecules inside the pipe to vibrate. The vibrating air inside the pipe creates a standing wave, which produces the beautiful resonating sounds typical of a pipe organ.

The different frequency sounds depend solely on the length of the organ pipe. The longer the pipe, the lower the frequency, whereas the shorter the pipe, the higher the frequency will be. Homemade organ pipes can be made by blowing into any glass or plastic bottle. The frequency of sound can be adjusted as well by simply adding or subtracting the length of the air column by adding or subtracting liquid. However, the quality of sound in the pipe does not necessarily depend on the length of the column, but instead depends on the materials and shape of the pipe.

Crystal wine glasses can shatter when resonating, but can they **make music**?

If the energy of a resonant standing wave is large enough, a crystal wine glass can shatter quite easily, but when the amplitude is smaller, the wine glass can produce a sound instead. Take, for example, a person rubbing his finger around the moist lip of a crystal wine glass. The glass seems to sing or hum. The humming is caused by the work the finger does on the glass: the rubbing of the glass causes it to achieve a standing wave pattern and perhaps resonate. The resonating molecules in the glass generate enough energy to vibrate the surrounding air and create a steady humming sound.

The frequency of the sound can be adjusted the same way that the frequency in an organ pipe can be adjusted—by changing the height of the air column, or in the case of the wine glass, by adding or taking away some of the liquid inside the glass.

IMPEDANCE

What is **impedance matching**?

When a wave travels from one medium into another, a percentage of the wave's energy travels into the new medium, but some of the energy reflects back into the original medium. In order to have a greater percentage of the wave's energy pass into the new medium, an impedance matching device needs to be introduced between the two media to allow for a smooth transition and to discourage any reflection.

How is impedance matching used in **shock absorbers**?

Shock absorbers are used to dampen the vibrations cars experience when they encounter bumps, pot holes, or other irregularities in a road surface. "Shocks" are designed to match the impedance of a vibration so the vibration does not cause the vehicle to repeatedly rock up and down. Instead of the vibration reflecting back and forth between the vehicle and the wheels, a piston in the shock absorber pushes its way into a fluid-filled cylinder. The fluid, typically oil (a fairly viscous fluid), absorbs most of the energy from the vibration, and reduces the up and down vibration dramatically. Effective shock absorbers should only allow the vehicle to bounce up and down a maximum of two times before all the energy is absorbed by the shocks.

What are **transformers**?

Transformers are used to match impedances when mechanical waves encounter new media. Instead of an abrupt barrier between the two media, a transformer provides a smoother, gradual transition from the old to the new medium. Depending upon the wave and the medium, different transformers, such as quarter-wavelength and tapered transformers, can be used to help minimize reflection. For example, electrical transformers are used to increase or decrease the current of electrical waves when entering different types of electrical devices. If a transformer is not available for a smooth transition, the impedance of the wave would not be matched and reflecting electrical waves would result.

An example of a tapered transformer can be found in sound-proof rooms or sound studios. Any sound that is produced is supposed to be absorbed by the impedance matching material on the walls. Special foam, tapered in V-like shapes, is used as a transformer to gradually absorb all the sound into the walls. The gradual changeover from the air medium to the wall medium prevents sound from reflecting back into the air.

An example of a quarter-wavelength transformer can be found on many camera lenses. The quarter-wavelength coating on a lens is used to bring the light wave into the lens without reflecting light back out into the air.

What is a **transducer**?

The purpose of a transducer is to change one type of wave into another type of wave. Telephone receivers are transducers that change longitudinal sound waves into electrical signals that pass through the telephone lines. Speakers are transducers that change transverse electric signals into longitudinal sound waves. Photoelectric cells are transducers that change solar electromagnetic waves into electrical signals.

THE DOPPLER EFFECT

What is the **Doppler effect**?

The Doppler effect is the change in frequency of a wave that results from an object's changing position relative to an observer. A well-known

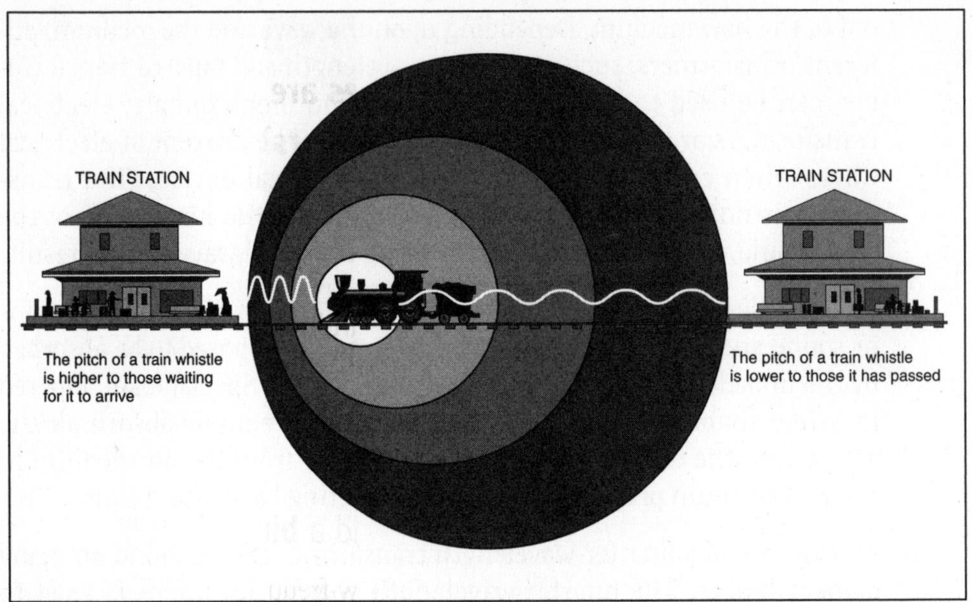
TRAIN STATION
TRAIN STATION
The pitch of a train whistle is higher to those waiting for it to arrive
The pitch of a train whistle is lower to those it has passed

The Doppler effect.

example of the Doppler effect is when a race car zooms by along the race track and makes a "wheee-yow" sound. The "whee" is caused by sound waves that are bunched together because the car is moving in the same direction as the emitting sound waves. The bunching together of sound waves creates an increase in the frequency and results in a higher-pitch sound. The "yow" sound occurs when the vehicle moves away from the propagation of the sound wave. Since the car moves away from the sound wave, the spacing between successive waves becomes greater. This decrease in the frequency of the sound wave results in a lower sound.

Who was the Doppler effect **named after**?

Johann Christian Doppler (1803–53), the Austrian physicist for whom the Doppler effect is named, discovered the change in frequency of a moving object while observing double stars. He found that the faster an object approaches or leaves a point, the greater the change in frequency. The knowledge gained from Doppler's observations has many uses in today's technological world.

What does the fact that most galaxies are seen with a red shift mean to astronomers?

The fact that astronomers observe most of the galaxies in the universe as having a red shift means that overall, galaxies are moving away from our galaxy, the Milky Way. This can only be happening if the universe as a whole is expanding, and is perhaps the greatest impetus to the development of the big bang theory.

What is the difference between a **red shift** and a **blue shift**?

The visible color spectrum ranges from the low-frequency red, orange, and yellow, to the higher-frequency green, blue, indigo, and violet. Astronomers observing the planets, stars, and galaxies use the Doppler effect to measure the speed at which objects are traveling, rotating, or revolving. For example, the rotational velocity of Saturn can be measured by observing the Doppler effect on the planet itself. As one section of Saturn turns toward the Earth, the other side of Saturn rotates away from the Earth. Therefore, the frequency of light emitted from the side moving toward the Earth gets higher, resulting in a blue shift. The light from the side spinning away from the Earth produces a lower frequency, called a red shift. Depending upon the speed of the planet, the color it emits varies in frequency as a result of the Doppler effect. This shifting or change in color, along with the intensity of the color, has allowed astronomers to determine the velocity of Saturn as approximately 11,000 kilometers per hour.

How do the police use the Doppler effect in **radar guns**?

The police use the Doppler effect when checking for speeding vehicles. A radar gun sends out radar waves at a particular frequency. As the radar wave hits a vehicle, the wave reflects back toward the radar gun at a different frequency. The frequency of the reflected wave depends upon how fast the vehicle is traveling. The faster the speed, the greater the frequency. Calculating the difference between the emitted frequency and the reflected wave's frequency, the radar gun determines the speed of the vehicle.

RADAR

What is **radar**?

Radar, a frequency band on the electromagnetic spectrum, is an acronym for "radio detection and ranging." Radar involves emitting electromagnetic waves and calculating the time, frequency, and directional changes of the reflected waves to locate the position and speed of an object. Radar is used in many different arenas, but was first used for military purposes to locate ships and planes when visibility was poor.

Who developed radar?

In 1935, Robert Watson-Watt, a Scottish physicist, created the first radar defense system for the British military. Although the British government originally asked for a device that would fry Nazi pilots in their cockpits, Watson-Watt explained that this was not possible, but a reliable early-warning signaling system might be feasible with the technology of the early 1930s. Watson-Watt used the information and breakthroughs from such physicists as Hertz and Marconi, the inventors of the first radio transmitter and antenna, to develop a British radar network that could detect enemy planes 100 miles off the coast of England.

Ironically, Watson-Watt became a deserving victim of his own technology nineteen years later. According to Canadian police, Watson-Watt had been speeding on a stretch of Canadian road, and was detected by a police radar gun. Watson-Watt willingly paid the fine and drove away.

What are some **other uses of radar technology** spurred by the military use of radar?

After its development for the military in World War II, the public realized the opportunities for the use of radar in everyday life. Weather forecasts have been improved by the technology found in today's NEXRAD Doppler radar systems. The skies have been much safer for airline travel and homes have been protected through the use of radar burglar alarms. Magnetic resonance imagining (MRI) uses radar technology to diagnose serious illnesses.

How has radar been used in **astronomy**?

Radar astronomy sends out radar waves and calculates position, velocity, and shape of objects in our solar system by analyzing the radar reflections. In the early 1960s, radar was used to determine the exact distance between the Earth and Venus. Later, radar was emitted from the space probe Magellan to map the surface of Venus. Radar astronomy has been beneficial in determining distances in our own solar system, but has not been used to measure distances outside our solar system.

NEXRAD DOPPLER RADAR

What is **NEXRAD Doppler radar**?

NEXRAD, or next-generation weather radar, is one of the most recent technological breakthroughs for weather forecasting. NEXRAD relies on the Doppler effect to calculate the position and the velocity of weather elements such as fronts, snow, rain, and dust particles. The spherical NEXRAD radar tower emits radar waves 360° around and calculates the frequency shift of the reflected radar waves off different weather elements. The NEXRAD computers then translate the information and represent the possible weather problems on a color-coded monitor for analysis.

The goal and main function of NEXRAD precision radar is to save American money and lives by predicting threatening weather problems and warning the public before tragedy strikes. Meteorologists already feel as though this new tool for weather forecasting has saved millions of dollars and many lives through its early warning systems. One of the most impressive advancements has been in pinpointing tornadoes more accurately than what was possible before NEXRAD.

What is the **cost and range** of NEXRAD radar?

Although many of the NEXRAD radar systems have been built and are in operation, there are still some of the 175 proposed NEXRAD systems that have yet to be completed. The cost for the entire project is estimated to be approximately half a billion dollars. Each NEXRAD station will

What do radio astronomers hear?

Radio astronomers measure noise patterns that pinpoint the location and characteristics of other galaxies, pulsars, and quasars. In order to listen to such signals, radio telescopes are used. These are shaped as large satellite dishes, and are able to detect wavelengths anywhere between 1 millimeter to 1 kilometer.

be able to accurately scan a radius of 125 miles, and less accurately up to 200 miles. Although the cost of NEXRAD is quite substantial, the lives that it is expected to save through early warning weather systems should be worth $500 million price tag.

RADIO ASTRONOMY

How is **radio astronomy** different from **radar astronomy**?

Radar astronomy measures the reflections of artificially emitted radio waves to determine an object's size, position, and velocity. Radio astronomy is a variation of radar astronomy in that it does not emit radio waves and wait for echoes or reflections; instead, it only listens for naturally occurring signals from other sources in the universe.

Where is the **largest radio telescope**?

The largest singular land-based stationary radio telescope is the 1000-foot diameter Arecibo Telescope, located at the Arecibo Observatory in Puerto Rico. It is continually pointed up to the heavens to pick up the noise from various locations.

A group of 27 radio telescopes, spaced 13 miles apart from each other in Socorro, New Mexico, have the ability to achieve extremely accurate

The Arecibo Observatory in Puerto Rico, home of the largest singular land-based stationary radio telescope.

readings. When combined, this group of radio telescopes, known as a Very Large Array (VLA) telescope, acts as a single telescope with a dish diameter of 13 miles. Such a large diameter is extremely helpful in seeing small details in radio astronomy.

Larger still are Very Long Baseline Interferometry radio telescopes. These radio telescopes are located throughout the United States (spaced much farther apart than VLA telescopes) and therefore have a diameter of thousands of miles.

SOUND WAVES

What is the **source** of every sound?

Sound waves are created by some type of mechanical vibration. Typically, sound originates from a vibrating object that forces the surrounding medium to vibrate. A tuning fork is an excellent example of a vibrating sound source. When struck by a rubber mallet, the tuning fork can be observed moving back and forth at its particular frequency. The vibrations cause the air molecules around them to move back and forth at the same frequency, creating areas of compressions (where the molecules are close together), and rarefactions (where the molecules are spread apart).

What type of wave is a **sound wave**?

A wave that consists of compressions and rarefactions—such as a sound wave—is called a longitudinal wave. As is the case with transverse waves, the medium or material that the wave is traveling through does not get transferred from sender to receiver; the molecules only vibrate back and forth from a fixed position.

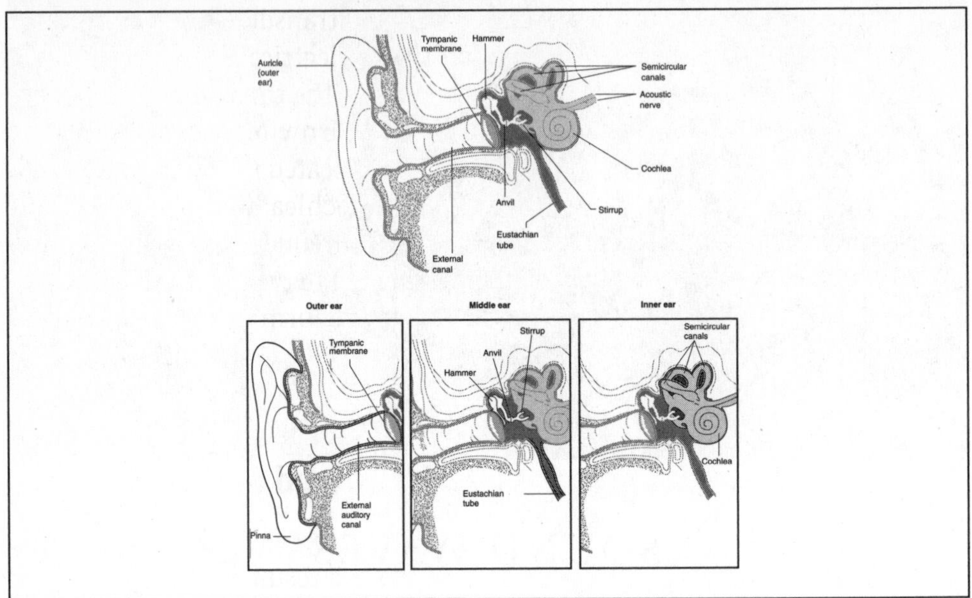

The human ear.

HEARING

How does a **person hear** sound?

The ear is the organ used to detect sound in humans and some animals. The ear consists of three major sections: the outer ear, middle ear, and inner ear. The external section of the ear consists of a cartilage flap called the pinna. The pinna's size and shape are used as a transformer to match the impedance of the sound wave entering the ear canal by gradually funneling the wave's sound energy into the ear. To hear more sound, people can increase the size of the pinna by cupping their hand around the back of the pinna—in effect, increasing its size and funneling capabilities.

Once the sound has entered the ear canal, it moves toward the ear drum, where the longitudinal waves cause the ear drum to push in or out depending upon the frequency and strength of the compressions and rarefactions. On the inner side of the ear drum lie the three smallest bones in the human body, the hammer, anvil, and stirrup. These three bones are connected to the ear drum and transfer the energy of the sound wave to the inner ear.

The function of the inner ear is to act as a transducer, by changing the longitudinal sound wave into a transverse electrical wave that is sent off to the brain for analysis. In order to do this, the three bones vibrate back and forth on the oval window, which in turn vibrates the fluid in the inner ear. The vibrating endolymph fluid, located in the cochlea, excites little hairs of varying lengths inside the cochlea; these hairs are called cilia. Depending upon the vibrating fluid's frequency, certain lengths of cilia will resonate at that frequency, sending nerve impulses to the auditory canal, which relays information in the form of electrical waves to the brain for analysis.

Why do people **hear ringing** after leaving a **loud rock concert**?

After leaving loud rock concerts, many concert-goers often complain of ringing in their ears. The ringing sound is a result of the destruction of the cilia by the high-volume sounds. Resonating objects often end up damaged or destroyed. Cilia resonate when a sound reaches the hair's natural frequency. If the sound is extremely loud and goes on for some period of time, it can cause destructive resonance to the cilia and kill it—the ringing sensation is actually the cilia dying. Usually the ringing is gone the day after the concert, but permanent damage has already been done, because those hair cells will never grow back. Although the effects of such hearing loss may take many years and repeated exposure to loud sounds to become apparent, they can nonetheless become very devastating.

What are the best ways to **protect one's ears** at **loud rock concerts**?

The first protection against damage to the cilia cells is to increase the distance from your ears to the speakers. The inverse square law dictates that the intensity of a sound is inversely proportional to the distance squared. In plain English, the farther away one is from the speakers, the lower the intensity of the sound. By simply doubling the distance, the intensity becomes one fourth of what it was originally.

The second method of protecting one's ears is to dampen the sound waves as they enter the ear. Many rock stars, after years and years of

Chris Cornell of the rock group Soundgarden points out his own ear plugs in concert (see previous page).

Why does one's voice sound different when heard on a recording?

How a person hears him or herself is unique to that person. When you speak, you hear yourself through sound waves propagating through your body, in addition to the waves propagating through the air. To make a sound, a person vibrates his vocal chords, which vibrate the different media around the vocal chords. These media are not only air, but tissue, bone, and cartilage as well. Waves travel through these media at varying speeds and create slightly different sounds when striking the ear. Thus, our voices on a recording sound funny to us because we are hearing them without the special characteristics they pick up when emanating from our bodies.

gradual hearing loss, now use ear plugs to decrease the amplitude of the wave entering the ear. The fluid in the cochlea transfers less energy to the cilia than if the listener were wearing no hearing protection.

SPEED OF SOUND

How **fast** does sound travel?

Light travels almost one million times faster than sound—specifically, 880,000 times the speed of sound. Light and all electromagnetic waves travel at a speed of 3×10^8 m/s (meters per second), while the speed of sound is only about 340 m/s or 760 mph (miles per hour) on a typical spring day.

The speed of sound compared to the speed of light can be observed at a baseball game. A spectator sitting in the outfield bleachers sees the bat-

Does sound travel faster on a hot or cold day?

Air molecules tend to move easier in hot and humid environments, due to their increased internal energy. Since sound relies on molecules bumping into one another to create compressions and rarefactions, the elasticity of molecules is what helps sound waves move faster. Therefore, on hot and humid days, sound travels faster than on cool, dry days when the air molecules are not as free to oscillate.

The following formula is used to determine the speed of sound in air:

Speed of Sound = 331 m/s + 0.6 (°C)

The speed of sound increases by 0.6 meters per second for every degree Celsius.

ter hit a ball before she hears the crack of the bat. The time delay for sound is quite large compared to that of light.

Who determined that sound needs a **medium** through which to travel?

In the 1660s, English scientist Robert Boyle proved that sound waves need to travel through a medium in order to transmit sound. To prove this, Boyle placed a bell inside a vacuum and showed that as the air was evacuated from the chamber, the sound of the bell became softer and softer, until there was no sound.

What did **Newton** add to the knowledge of sound media?

Although he mainly concentrated on classical mechanics and the principles of geometric optics, Newton did make several important discoveries in the field of sound. His major contribution was his work on sound wave propagation. He showed that the velocity of sound through any medium depended upon the characteristics of that particular medium. Specifically, Newton demonstrated that the elasticity and the density of the medium determined how fast a sound wave would travel.

How **fast** does sound travel in **different media**?

The speed of sound in a medium depends upon several factors, such as density, temperature, whether the medium is a liquid or solid, and the elasticity of the medium. The more elastic the medium, the faster the sound wave will travel. The following table illustrates some examples of how fast sound can travel in different media.

Medium	Speed of Sound (meters/second)
Air (0°C)	331
Air (20°C)	343
Air (100°C)	366
Helium (0°C)	965
Mercury	1452
Water (20°C)	1482
Lead	1960
Wood (oak)	3850
Iron	5000
Copper	5010
Glass	5640
Steel	5960

How could sound be used to determine if **global warming** has taken place?

The ATOC, or Acoustical Thermometry of Ocean Climate, has proposed a controversial experiment that would help determine the extent to which global warming has taken place. The ATOC experiment argues that atmospheric global warming has only increased by half of what original studies predicted. Many climatologists feel that the oceans' absorption of heat is to blame for the subtle increase in atmospheric temperature. In order to prove this theory, the temperature of the oceans must be mea-

Is there a way to figure out how far away lightning has struck?

Lightning and thunder occur at the same time. However, light travels approximately 880,000 times faster than sound on a typical day. Although it takes virtually no time to see lightning, depending upon the observer's distance from the lightning, it can take quite a while to hear the thunder.

From the differential between the two speeds, a general rule has emerged to determine how far away lightning has struck. Upon witnessing lightning, count the number of seconds it takes before hearing thunder. Since the sound of thunder travels slower than light, divide the number of seconds between the lightning and thunder by five to determine how many miles away the thunder and lightning occurred. For example, if you see a flash of lightning and approximately 10 seconds later you hear the thunder, divide the 10 seconds by the number 5 to find that the lightning occurred 2 miles away.

sured to see if they have indeed absorbed heat from the atmosphere and become warmer due to the greenhouse effect.

To measure the temperature of the ocean, the ATOC proposed placing large loudspeakers on the ocean floor that would produce loud, repeated bursts of 75 Hz frequencies for 20-minute intervals. Receivers, attached to a central computer, would be placed on other sides of the ocean to receive the signals. The computer would then calculate how long it took the sound to travel from Hawaii or California, the speaker sites, to the receivers scattered between New Zealand and Alaska. By measuring the time, scientists would be able to determine if the water has become warmer or cooler over the life of the experiment.

Sound waves are an effective method of measuring the temperature of water for several reasons. First of all, sound waves vary their speed in accordance with the medium in which they are traveling. The warmer the medium, the faster the sound travels. In fact, sound traveling through water moves a full 4.6 meters per second faster per degree Celsius. By determining the time the sound takes to travel a particular distance, scientists can calculate how fast the sound travels and thus, the temperature of the water. Secondly, sending sound waves through water is an effective method for measuring the temperature because sound does not reduce its amplitude in water as easily as it does in air. This results in a reliable, efficient, and unique way of testing for global warming. However, this plan is controversial because some scientists feel that it is perhaps harmful to subject aquatic animals to sounds that we humans may not be able to hear, but that the aquatic animals might be able to hear. The bombardment of low-frequency sounds such as proposed by the ATOC could have a psychological effect on the animals.

ULTRASONICS AND INFRASONICS

What are the **frequency limits** of the human ear?

The human ear's anatomy allows humans to hear frequency ranges between 20 Hz (hertz) and 20,000 Hz. The lower and upper fringes of this bandwidth can be difficult to hear, but many people—especially younger people—can hear these frequencies quite well.

What frequencies can our ears **most readily detect**?

The typical human ear can best detect frequencies between 200 Hz and 2000 Hz. Although humans hear the other sections of the human bandwidth, human ears are designed to be most sensitive to these frequencies.

What are the **bandwidths of hearing** for other animals?

Animal	Lowest Frequency (Hz)	Highest Frequency (Hz)
Human	20 Hz	20,000 Hz
Dog	20 Hz	40,000 Hz

Animal	Lowest Frequency (Hz)	Highest Frequency (Hz)
Cat	80 Hz	60,000 Hz
Bat	10 Hz	110,000 Hz
Dolphin	110 Hz	130,000 Hz

ULTRASONICS

What are **ultrasonic** sounds?

Ultrasonic sounds are those frequencies above the human bandwidth of hearing. Frequencies above 20,000 Hz are not heard by people, but do exist in the environment. Other animals are quite sensitive to ultrasonic frequencies: dolphins, for example, use ultrasonic frequencies to communicate; bats use ultrasonic sounds as a tool for navigation and hunting.

What is **sonar**?

Sonar, an acronym for "sound navigation ranging," is a method of using sound waves to determine the distance an object is from a transmitter of sound. These sound waves (usually clicks of ultrasonic sound), are emitted from a transmitter, reflected off an object, and reflected back to the receiver section of the transmitter. The device measures the length of time it took for the sound wave's round trip, and uses the speed of sound to calculate the distance the object is from the transmitter.

Sonar is used (predominantly as a navigational tool) by humans and animals alike. Machines such as depth-finders on boats, stud-finders and range meters in construction, and motion detectors for security devices all employ sonar. Dolphins and bats, among other animals, use sonar for navigation, hunting, and communicating.

What is **ultrasound**?

Ultrasound is a method of looking inside a person's body to examine tissue- and liquid-based organs and systems without physically entering the body. Ultrasound systems direct high-frequency sound (usually between 5 and 7 megahertz) into particular regions of the body, and

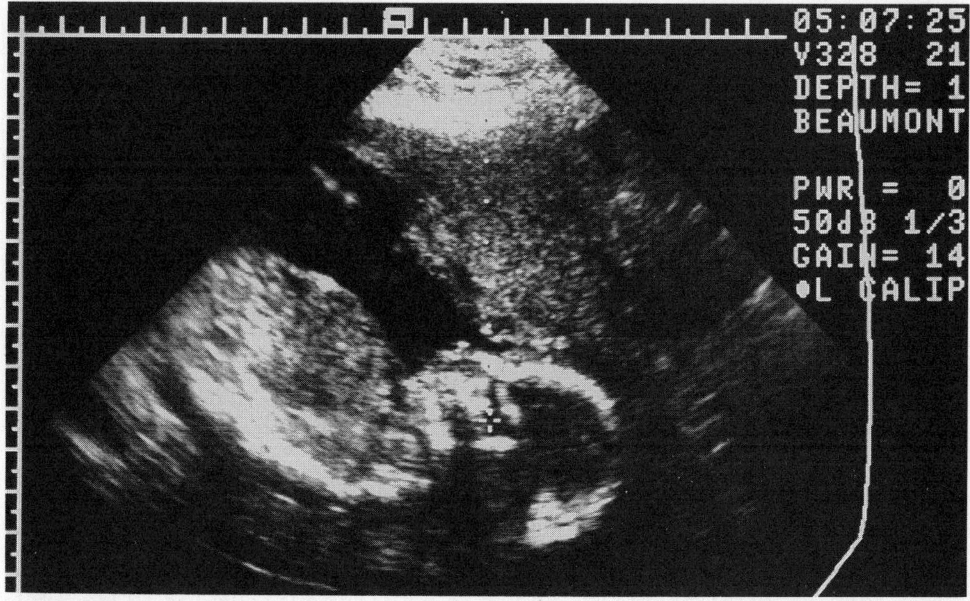

Ultrasound of a fetus.

measure the time it takes for the sound wave to reflect back to the machine. By analyzing the pattern of reflections received, a computer can create a visual representation on a monitor.

Ultrasound is sometimes used instead of X-ray, for it does not employ radiation and is safer for the person being examined. Obstetricians use ultrasound to examine the progress and/or problems that a fetus might be experiencing. This method is also used to observe the status of different fluid-like organs and systems in the body such as the nervous, circulatory, urinary, and reproductive systems. Ultrasound is not used to examine bone structures, for the ultrasonic sound waves are absorbed by skeletal structures—they only reflect off of liquid-based objects.

INFRASONICS

What are **infrasonic** sounds?

Whereas ultrasonic sounds are frequencies *above* the human bandwidth of hearing, infrasonic (or "subsonic") sounds are those frequencies *below*

Can infrasonic sounds be dangerous?

There have been studies suggesting that bridges vibrating at infrasonic frequencies could perhaps encourage heart dysrhythmia in people living next to them. The low-frequency sound waves emitted from the bridge are not audible to humans, but it is speculated that the waves could perhaps affect the heart. This has not been proven, however.

the human bandwidth of hearing. Infrasonic sounds, usually created by slow vibrating objects such as bridges, produce sound frequencies less than 20 Hz. Elephants are known to make sounds as low as 12 Hz, while nuclear explosions can produce infrasonic sounds as low as 0.01 Hz.

How can infrasonic sounds provide early warning of **tornadoes**?

Using sound sensors that detect infrasonic sound frequencies, scientists discovered quite by accident that the spinning core or vortex of a tornado produces sounds that are a few hertz below the human bandwidth of hearing. The tornado, much like the organ pipe, produces low frequencies when the vortices are large, and higher frequencies when the vortices are small. Since the infrasonic sound waves from tornadoes can be detected for up to a 100 miles away, they could help increase the warning time for tornado strikes.

INTENSITY OF SOUND

What is **sound intensity**?

Sound intensity is the energy of the sound wave. For sound and all mechanical waves, the energy is determined by the height of its amplitude. The amplitude of a sound wave represents its intensity or volume—the larger the amplitude, the louder the sound.

What does the **pitch** of a sound mean?

The pitch of a sound is the combination of frequency and intensity. A high-frequency sound that is played loudly will sound higher in pitch than the same frequency played at a lower volume. A low-frequency sound that is played loudly will sound lower in pitch than the same frequency played at a lower volume.

What does the **Doppler effect** have to do with pitch?

The Doppler effect demonstrates that a wave's length decreases as the source moves toward an observer, and increases when the source moves away. Specifically with sound, not only does the wavelength change when the source is moving, but the intensity—or pitch—of the sound changes as well. The closer the object emitting the sound, the greater the intensity or amplitude of the sound. As the object moves away, the intensity decreases.

Why does sound diminish as its source **moves away**?

There are two major answers to this question. First, the amplitude of a sound's wave or its intensity would be dampened in its travels by friction between the air molecules. The second reason for the decrease in intensity is because sound waves do not travel in a narrow path, but spread out into the surrounding medium in a spherical form. (It would be quite difficult to hear something if the sound did not spread out, for we would have to aim our sound waves directly at another person's ear in order for them to receive the message.) The amount of energy in the emitted sound is fixed, so the energy per unit area decreases as the area increases.

How much does a sound's intensity **diminish** as it moves away?

Sound intensity diminishes according to the inverse square law, which states that the intensity of a sound is inversely proportional to the square of the distance. For example, if a person stands 1 meter away from a speaker, the sound intensity might be an arbitrary unit of 1. If that same person moved so she was 2 meters away from the speaker, the intensity would be 1 over the square of the distance, or 1/4th the intensity. Again, if the listener moved 3 meters away, the intensity would be 1/9th the intensity it was at the 1-meter mark.

What is a **decibel**?

A decibel (abbreviated dB) is the internationally adopted unit for the relative intensity of sound. The intensity is "relative" because the measurement compares a loudness level to a reference level, usually the threshold of human hearing. The decibel scale is a logarithmic scale, meaning for every 10 decibels, the loudness of a sound is increased by a factor of ten. For example, a relative intensity change of 30 dBs to 40 dBs means the sound will be 10 times louder than it was at 30 dBs. A change of 30 dBs to 50 dBs would mean the new sound would be 100 times as loud.

The following chart shows a typical sound environment, how many times louder those levels are than the threshold of human hearing, and the relative intensity of that sound compared to the threshold of hearing.

Sound Environment	Times Louder than Threshold	Relative Intensity (decibels)
Loss of hearing	1,000,000,000,000,000	150
Rocket launch	100,000,000,000,000	140
Jet engine 50 meters away	10,000,000,000,000	130
Threshold of pain	1,000,000,000,000	120
Rock concert	100,000,000,000	110
Lawnmower	10,000,000,000	100
Factory	1,000,000,000	90
Motorcycle	100,000,000	80
Automobiles driving by	10,000,000	70
Vacuum cleaner	1,000,000	60
Normal speech	100,000	50
Library	10,000	40
Close whisper	1,000	30
Leaves rustling in the wind	100	20
Breathing/whisper 5 meters away	10	10
Threshold of hearing	0	0

What is the maximum decibel level that a person can experience without pain?

The threshold of pain for humans depends on the person in question, but typically ranges between 120 dBs and 130 dBs. Such pain can be experienced at extremely loud rock concerts and next to jet engines and jackhammers, for example.

Are there **federal standards** for using **hearing protection**?

Federal regulations mandating the use of hearing protection in the workplace state that if an employee works for eight hours in an environment in which the average noise level is above 90 decibels, the employer is required to provide free hearing protection to those employees. For example, many high school and college students work for landscapers in the summer. Since the average decibel level for a lawnmower is about 100 dBs, and if the employees are working for eight or more hours per day, free ear plugs or ear muffs must be provided to the employees.

The Sound Barrier

See also: FLUIDS chapter, "FLUID DYNAMICS"

What is the **sound barrier**?

The sound barrier is the speed that an object must travel to exceed the speed of sound. The speed of sound is often used as a reference with which to measure the velocity of an aircraft. The speed of sound, approximately 331 m/s (meters per second) at 0°C, is considered Mach 1. Twice the speed of sound is Mach 2, three times the speed of sound is Mach 3, and so on.

What was the **first plane** to break the sound barrier?

The first plane to break the sound barrier was the Bell X-1, piloted by Chuck Yeager in 1947. The plane was launched from another plane high in the atmosphere, and after firing its rocket engine, the X-1 achieved a speed of 1,229 kilometers per hour (701 miles per hour). Many thought it impossible to travel faster than the speed of sound, but today many jets travel several times the speed of sound.

What is a **sonic boom**?

A sonic boom occurs when an object travels faster than the speed of sound. The boom itself is caused by an object, such as a supersonic airplane, traveling faster than the sound waves themselves can travel. The resulting compression of the sound waves creates the "boom" when striking a person's ears. A sonic boom is not a momentary event that occurs as the plane breaks the sound barrier; rather, it is a continuous sound caused by a plane as it travels at such a speed.

All objects that exceed the speed of sound create sonic booms. For example, missiles and bullets, which travel faster than the sound barrier, produce sonic booms as they move through the atmosphere. If, however, the planes, missiles, and bullets were in space, there would be no sonic boom or sound, due to the absence of air, or any medium, for that matter.

ACOUSTICS

What is **acoustics**?

Acoustics is the branch of physics that deals with the science of sound. Although sound has been studied in earnest since Galileo made some predictions in the early 1600s, the ability to study sound has grown tremendously since the advent of electronic measuring devices and generators such as graphic equalizers, synthesizers, and various recording mechanisms.

What is an **acoustical engineer**?

Acoustical engineers or architects design and construct environments where sound is intended to be pleasing to the ear, such as amphitheaters, concert halls, auditoriums, sound and radio studios, sound-proof rooms, highway noise barriers, and more. Acoustical engineers and architects must take into consideration the design and materials that will make a listening environment pleasing to an audience. The goal of acoustical engineers is to create areas that cause just enough reflection and damping of sound waves to make the voice, music, or sound seem natural.

What is **reverberation time**?

The reverberation time for a sound is the time it takes for the echoing sound to diminish to 1/1,000,000th of its original intensity. In other words, the reverberation time is how long it takes for the sound to diminish to an amplitude that our ears cannot detect. The longer the reverberation time, the more echoing is heard because the sound has more time to reflect off walls and other surfaces. The shorter the reverberation time, the less echo is heard.

What role does **reverberation time** play in **acoustics**?

Reverberation time plays a major role in the quality of sound heard in a concert hall or sound studio. Acoustical engineers carefully design concert halls to achieve a typical reverberation time between one and two seconds. If a reverberation time is too short, as in sound-proof rooms, the sound will diminish almost instantaneously and lack the fullness that most pleasing sounds possess. If the reverberation time is too long, as it is in many gymnasiums, the echoing effects will interfere with the new sounds, making the words or sounds difficult to understand.

What is **percent reflection**?

The percent reflection for a sound is the amount of sound a material absorbs versus the amount of sound it reflects back into the environment from which the sound emanated. Good absorbers of sound match

What materials are effective absorbers of sound?

Different materials will absorb certain frequencies of sound better than other frequencies, but some of the best absorbers of sound are soft objects. Materials such as felt, carpeting, drapes, foam, and cork are good at matching the impedance of a sound wave and reflecting back very little sound energy. However, materials such as concrete, brick, ceramic tile, and metals are effective reflecting materials of sound. That is why gymnasiums (with hardwood floors, concrete walls, and metal ceilings) have relatively long reverberation times, while concert halls furnished with cushy seats, carpeted floors, and long drapes have relatively short reverberation times. Simply the materials alone can create a particular listening environment based on its percent reflection.

the impedance of sound waves by transforming the sound into the new medium, while poor absorbers of sound wave energy do not match the impedance and reflect the sound back into the environment. The equation for percent reflection is as follows:

Percent reflection = (energy reflected / initial energy) × 100%

What was the first **concert hall** to be designed by an acoustical physicist?

The Boston Symphony Hall, designed by physicist Wallace Sabine, is considered to be the first concert hall designed specifically to enhance the sound of an orchestra. Sabine, who designed the hall in the late 1890s, discovered the relationship between sound absorption, reverberation time, and sound intensity. Sound reflections can either enhance or ruin a sound. Sabine discovered that generating strong reflections immediately after a sound was produced would enhance sound, whereas if a sound bounced off an object later in its propagation, it would detract from the sound because it would be slightly out of phase with the first sound wave (the basis for destructive interference).

Sabine's Boston Symphony Hall, built in 1900, established an excellent reputation for sound quality, mostly due to the choices of sound absorption material as well as the strategic placement of reflecting material. The goal was to use the sound reflecting materials (high percent reflection ratios) to create strong initial reflections, while using sound absorbing materials (low percent reflection ratios) to absorb most of the energy from sound that would ordinarily reflect off of the high ceiling and the side walls in the rear of the hall.

Why are some concert halls shaped like **funnels**—narrow up front and wide toward the rear?

The reason that concert halls tend to spread out toward the rear is to provide the hall with a megaphone effect. Megaphones are used by movie directors, cheerleaders, and protesters to increase the energy of sound waves. Concert halls employ the basic shape of a megaphone, called a tapered transformer, to contain the sound, not allowing it to immediately spread out in all directions, which increases the intensity of the sound by limiting the area over which the sound must—or can—spread.

What are **fundamental frequencies**?

A fundamental frequency is the lowest and most intense frequency that a particular sound makes. Although a trumpet and a French horn can both play the note middle C, which has a fundamental frequency of 256 Hz, this note doesn't sound the same on both instruments. What distinguishes a C on a trumpet from a C on a French horn is dependent on other frequencies, called overtones.

What are **overtones**?

Overtones are multiples of a fundamental frequency. A 256-Hz frequency produced on a trumpet will have a different intensity and number of overtones than a tuba or French horn; however, the placement of the overtones will be the same. The first overtone for a 256 Hz frequency would be at 512 Hz, the second overtone would be 768 Hz, the third overtone would be three times 256 Hz, and so on.

What is the difference between an **overtone** and a **harmonic**?

Overtones and harmonics are both terms that define the multiple frequencies of the fundamental, but are used in different contexts. "Overtone" is a scientific term used specifically in acoustical physics, whereas "harmonic" is a musical term. The fundamental frequency is the first harmonic, the first overtone is the second harmonic, the second overtone, the third harmonic, and so on.

What other factors contribute to the **quality of a sound**?

The number of overtones, as well as their frequency and intensity, affect the quality of sound that a person, instrument, or other sound-making device creates. As a general rule, the more overtones an instrument has, the more frequencies the instrument or person produces, resulting in a better quality sound, because a pure tone with no overtones (like a tuning fork) does not have much quality, musically speaking, whereas a saxophone, with many overtones, has a good-quality sound.

Who was Jean Baptiste Joseph **Fourier**?

Fourier, principally known for his trigonometric Fourier series equations, was a mathematician who established a mathematical system for overtones and harmonics. This allowed mathematicians and physicists alike to study sound in a quantitative manner. Later, musicians would see the benefits of Fourier's work on harmonics and use his system to help them analyze and create musical scores.

What is **filtered music or sound**?

Filtered sound has gone through a device to eliminate particular ranges of unwanted frequencies, such as high-frequency hissing, or low-frequency humming. High-frequency filters eliminate high frequencies, whereas low-frequency filters eliminate the lower frequency sounds. Filtering out low and high-frequency sound is sometimes used to eliminate distracting noise. What's left are frequencies that are within the desired range.

What are **difference tones**?

Difference tones are frequencies that are produced as a result of two different frequencies interfering with each other. An old-fashioned British police whistle is an example of a device that can create difference tones; they have two pipes and produce one high-frequency tone and one slighter lower. Instead of hearing just the two high frequencies, the interfering waves produce a third frequency, or difference tone, at a frequency that is the difference between the two original frequencies. For example, if the high-frequency sound was 812 Hz, while the lower frequency was 756 Hz, the difference tone from the interfering sound waves would be 144 Hz.

Why is it that sometimes a singing duet can sound as if there were a third voice contributing to the music?

When two people sing at slightly different frequencies, the frequencies can interfere, causing a difference tone, or third frequency (see above question). You would also experience this phenomenon with volunteer fire whistles blaring through a town, or even the beeps or tones emitted by a clock radio. In fact, many difference tones are intentionally created to enhance the sound.

NOISE POLLUTION

See also: "HEARING"

Is that crazy racket really **music**, or just **noise**?

Music versus noise is a relative question, depending upon a person's particular tastes, but for scientists, music represents a reproducible, distinct waveform, while noise consists of all types of sound interfering with each other, and as a result, possesses little usable or no recognizable information.

Why is noise pollution **dangerous**?

In the past, noise pollution was only thought to create health effects if the intensity was large enough to cause hearing damage. However, studies over the past several decades have found that long-term exposure to noise can cause potentially severe health problems—in addition to hearing loss—especially for young children. Constant levels of noise (even at low levels) can be enough to cause stress, which can lead to high blood pressure, insomnia, psychiatric problems, and can even impact memory and thinking skills in children. In a German study, scientists found that children living near the Munich Airport had higher levels of stress, which impaired their ability to learn, while children living further from the airport did not seem to experience the same problem.

What **limits** have been established to **reduce exposure** to noise pollution?

The World Health Organization has recommended that noise during sleep be limited to a level of 35 dBs (decibels), and governments are beginning to place restrictions on noise levels in both residential and business environments. In the Netherlands, for example, regulations specify that new homes may not be built in areas of high noise levels—those that exceed average noise levels of 50 dBs. In the United States, employers must provide hearing protection for those who endure noise levels of 90 dBs for more than eight hours a day.

What **methods** are being used to **reduce noise pollution**?

Since noise creates stress and can lead to other health problems, industries and governments around the world are working to reduce noise levels, especially around populated regions. One method of reducing noise pollution around airports has been rerouting airline traffic so that it passes over less-populated areas. New technologies, such as active

What is anti-noise and active noise cancellation?

Active noise cancellation is the name given to a relatively new technology designed to combat noise pollution. ANC, as it is often abbreviated, uses a microphone to receive noise, sends the signal to a microprocessor, reproduces the opposite wave form of the noise, and emits that opposite noise, or anti-noise, from a speaker. The anti-noise uses the principle of destructive interference to cancel out the original noise. As a result, the person listening to the anti-noise hears neither the original noise or the anti-noise.

noise cancellation, will help reduce and perhaps eliminate low-frequency noise levels in airplane jets, truck mufflers, and helicopter rotors. Sound barriers have been installed along many highways to absorb and/or reflect sound away from houses built alongside the roads. In countries such as Austria and Belgium, roadways are being constructed with a material called whisper concrete, which engineers claim reduces noise by 5 dBs (decibels). Finally, Swedish engineers have developed a road surface made of pulverized rubber that can reduce the noise level by as much as 10 dBs.

What are some of the **problems** with ANC technology?

The difficulty in using active noise cancellation is in creating the opposite waveform of the noise in order to cancel out the original noise; because noise is not considered a repeatable wave pattern, technology had to be created to predict what the noise would be, or to produce anti-noise that would cancel out repeating elements in the noise. The repeating element that is relatively easy to predict is the low-level frequency section of noise. As a result, low-frequency noises found in helicopter, jet engine, and muffler noise can be destructively interfered, while the high squealing sounds of jet engine noise are more difficult to cancel out.

If **anti-noise** is emitted by a speaker, can a person hear other people, music, etc.,—sounds that aren't "noise"?

The objective of active noise cancellation is to cancel out the noise waveform by producing anti-noise to interfere with the original noise pattern. Since the result is less noise, other sounds should be easier to hear. ANC technology is now available for consumers in headsets, that reduces noise and allows for easier communication in loud environments. As a result of this new technology, factory workers, helicopter pilots, and airline passengers should be able to communicate more easily, and their amount of stress due to noise pollution should be reduced.

Does my neighbor's **motorcycle** have to be that loud?

At times, engineers try to achieve just the right sound or noise for a particular product. The product, whether it's a vacuum cleaner, lawnmower, or motorcycle, needs to be quiet enough so as not to cause stress, yet has to have enough sound to seem powerful. For example, muffler technology has the ability to greatly reduce the noise a motorcycle produces, yet many engineers and manufacturers feel that consumers would not purchase the product if it does not sound "powerful" enough.

What is **psychoacoustics**?

Psychoacoustics, which connects acoustical physics with psychology, is the study of how the mind reacts to different sounds. This field of study is especially important to consumer product manufacturing, because a consumer associates particular sounds with certain products or sensations. For example, people associate low-frequency rumbling sounds with power and torque, while higher-frequency sounds oftentimes represent high speeds and out-of-control occurrences. Psychoacoustics can play a major role in the development and commercial success of many products.

PROPERTIES OF LIGHT

What is **light**?

Light, or visible light, as it is often called, is what enables vision. In fact, light is the only thing people see; every object we think we see is a result of light that is reflected off the objects and into our eyes—in the absence of light, we would see nothing.

Light is an electromagnetic wave, located between infrared and ultraviolet on the electromagnetic spectrum. Although light is on a wave spectrum, at times it mimics the behavior of particles. (The true nature of light, whether it is a wave or particle, will be discussed in the MODERN PHYSICS chapter.)

What **isn't** light?

Some ancient Greeks felt that invisible "streamers" emitted from our eyes were responsible for sight, while others felt that light consisted of particles flying through the air and striking our eyes. It was not until the time of Isaac Newton and Christian Huygens, two pioneers in the study of light and optics, that true scientific theories were formed concerning light.

THE VISIBLE LIGHT SPECTRUM

Where does visible light lie on the **electromagnetic spectrum**?

Visible light is an extremely narrow band on the electromagnetic spectrum. The low-frequency light is just a bit higher than the infrared section at 4×10^{14} Hz, while the higher-end frequencies of visible light end at 7.9×10^{14} Hz. The entire section of visible light, when combined together, is seen as white light. However, if only a small section of the visible light spectrum is seen, the light appears as a particular non-white color.

THE SPEED OF LIGHT

How **fast** does light travel?

Because light is found on the electromagnetic spectrum, light travels at the same speed as other electromagnetic waves, which, in a vacuum, is 299,792,458 m/s (meters per second), or 186,282.4 mi/sec (miles per second). Except when performing experiments that need the exact velocity of light, this number is usually rounded up to 3×10^8 m/s (186,000 mi/s). Just like any other wave, light does slow down when it enters the Earth's atmosphere.

Can the speed of light **change**?

Waves travel at different speeds when traveling in different materials. In the vacuum of space, light travels at 3×10^8 m/s (meters per second). If light could travel in a circular path, it would orbit the Earth 7.5 times in one second. When light encounters a denser medium, however, like that of the Earth's atmosphere, it slows down ever so slightly to 2.91×10^8 m/s. Upon striking water it slows down rather dramatically, to 2.25×10^8 m/s, three quarters of its original speed. Finally, when light passes through the dense medium of glass, it slows to only 1.98×10^8 m/s. Even at this slower speed, light would still travel around the Earth (if it could indeed travel in a circle) 5.6 times in one second.

Who is Roy G. Biv?

The colors of the visible light spectrum, in order from low frequency to high frequency, can be remembered by the fictitious name "Roy G. Biv," which is an acronym for red, orange, yellow, green, blue, indigo, and violet. When combined, these colors form white light.

Each individual color has a particular frequency, and as the frequency changes, the color begins to take on the characteristics of colors next to it on the spectrum.

Who was the first physicist to make a serious effort to **measure** the speed of light?

Light travels at a very high speed and therefore can be extremely difficult to measure. Prior to 1600, most people believed that light traveled instantaneously. However, Galileo felt that there had to be some finite speed of light and attempted to measure it by gauging the amount of time it took for a distant light to reach his eyes. To perform this experiment, Galileo had an assistant with a lamp stand a great distance away from him. Galileo instructed the assistant to uncover his lamp immediately when the helper saw Galileo uncover his own lamp. By measuring the time it would take the light to travel from Galileo and back again from the assistant, Galileo felt that he could measure the speed of light. His experiment failed because there was no possible way for him to measure such a short period of time. Galileo walked away from the experiment without a number for the speed of light, but with a deep appreciation for how fast light actually does travel.

What other significant attempts have been made to measure the speed of light more **accurately**?

Instead of using astronomical measurements to determine a speed for light, French scientist Armand Fizeau attempted to determine the speed

What astronomical attempts have been made to measure the speed of light?

Several astronomical observations and measurements in the late seventeenth century helped determine a speed for light. Physicist Christian Huygens, famous for his belief in the wave-like nature of light, calculated the speed of light using astronomical observations from Danish astronomer Olaus Roemer. Roemer discovered a discrepancy in the amount of time it took one of Jupiter's moons, Io, to orbit Jupiter. Sometimes Io seemed to orbit Jupiter rapidly, while other times it seemed to take as much as twenty-two minutes longer than its fastest orbit.

Huygens came to a conclusion after studying the positions of Jupiter and Io in relation to the position of the Earth: he said it was not the speed of Io that changed, but the distance between the Earth and Io that varied, and that if the Earth were farther away, it would take the light reflecting off Io longer to reach the Earth. By using Roemer's observations, Huygens calculated a speed of light of 2.2×10^8 m/s (meters per second).

of light in a laboratory environment in the mid-1800s. Fizeau used a series of mirrors, a rotating stroboscope (a wheel with little notches along the outer perimeter), and a light source to get a more precise measurement of the speed of light. By timing the rotation of the notched stroboscope and measuring the distance between mirrors, Fizeau calculated the speed of light to be 3.13×10^8 m/s (meters per second), a bit closer to the actual speed of light than Huygens' earlier calculations.

It was not until Jean Foucault, and later, American physicist Albert Michelson, that the speed of light was redefined once again. Michelson's experiment (which was exactly like Foucault's experiment, only a bit more accurate) was on a grander scale than Fizeau's, for it consisted of a light source, a rotating mirror, and a plane mirror located on Mount San

Antonio and Mount Wilson in California, 35 kilometers apart from each other. By measuring the speed of the rotating mirror and the distance between the mirrors, Michelson came up with the most accurate measurement of light to that date, a feat which, in 1907, gave him the honor of being the first American to win the Nobel prize in physics. After winning the Nobel prize, Michelson continued to work on a more accurate measurement for the speed of light, and in 1926 calculated it to be 2.997996×10^8 m/s, extremely close to today's accepted value.

What is used to **measure** the speed of light?

The advent of laser technology in the 1960s provided physicists with a new tool for measuring the speed of light. In the early '80s, the International Committee on Weights and Measurements designated the speed of light to be represented by the letter "c" with a value of 299,792,458 m/s (meters per second) in a vacuum. The speed of light has great importance, for it is now used as a world standard for measuring distance. For example, the international standard for the length of a meter is the distance it takes light, in a vacuum, to travel 1/299,792,458th of a second.

How long does it take **light** to travel...

one mile?	5.3×10^{-6} seconds
from New York to Los Angeles?	0.016 seconds
around the equator of the Earth?	0.133 seconds
from the moon to the Earth?	1.29 seconds
from the sun to the Earth?	8 minutes
from Alpha Centuri (next closest star)?	4 years

(Some of the destinations above would require light to travel in a circular or curved path, which it does not do under normal conditions.)

What field of science uses the light-year as a **measurement device**?

Astronomers deal predominantly with the study of electromagnetic waves, all of which travel at the speed of light. Therefore, to use the

What is a light-year?

The concept of a light-year can be rather confusing at first. A year is a unit of time, but a light-year represents a unit of distance. Specifically, it is the distance that light travels in one year. Since light travels at 3×10^8 m/s (meters per second) and a year consists of 31.536×10^6 seconds, a light-year is therefore equivalent to 9.46×10^{15} meters.

light-year as a unit of distance not only makes the mathematical calculations easier, but also makes conceptual sense for the astronomer as well. In addition, astronomers are more likely to *need* a unit of measurement that large, than, say, a biologist.

How are **amounts** of light measured?

The amount of light an object produces, either through generated light or reflected light, is measured in lumens. Specifically, lumens indicate how much light is emitted from an object; the unit of measurement for the intensity of light, however, is the candela.

As the distance between the source and receiver increases, the intensity of the light decreases according to the inverse square law: given a light 1 meter away with an intensity of 1 candela, moving the source twice as far away would result in the intensity being equal to 1 over the distance squared, or $1/2^2$, or 1/4 the intensity at the 1-meter mark. If the source were moved three times farther away, to 3 meters, the intensity would decrease to 1/9th the original intensity.

OPAQUE, TRANSPARENT, AND TRANSLUCENT OBJECTS

What is an **opaque** object?

An opaque object is something that allows no light through it. Concrete, wood, and metal are some examples of opaque materials. Some materials can be opaque to light, yet translucent to other types of electromagnetic waves. For example, wood does not allow visible light to pass through it, but will allow other types of electromagnetic waves, such as microwaves and radio waves. The physical characteristics of the material determine what type of electromagnetic energy will and will not pass through it.

What is the difference between **transparent** and **translucent**?

Transparent media such as air, water, glass, and clear plastic allow light and images to clearly pass through the material. Translucent materials, on the other hand, allow light to pass through, but clear images cannot be seen through them. For example, frosted glass and thin paper are translucent because they let some light through, but are not transparent because you cannot see clearly through them.

Why is it so important to have **ozone** in our atmosphere?

Ozone is semi-transparent to ultraviolet light, and helps protect us from harmful levels of ultraviolet (UV) light that we would otherwise be exposed to. Those of us who frequent the beach know what ultraviolet light can do to our skin when lying out too long in the sun. The recent concern over the decline in ozone in our atmosphere is quite serious and warranted. If large and widespread gaps in the ozone layer were to form, the ability to reflect the UV rays would be severely diminished. Physicists, meteorologists, and biologists have suggested that high exposure levels of UV light could prove harmful to humans and others animals and plants.

Shadows

What are **shadows**?

Shadows are simply areas of darkness created by an object blocking light. Whether created when someone puts their hand in the light from a movie projector, stands outside in the sunshine, or sees the moon move between the Earth and the sun during an eclipse, shadows have always intrigued us.

How is an **eclipse** a shadow?

An eclipse is created just as any other shadow is formed, by the presence of an object in the path of light. In an eclipse, the object blocking the light would either be the Earth or the moon moving in front of the sun. The blockage of the sun's light forms the shadow, called an eclipse.

Eclipses

What is a **lunar eclipse**?

A lunar eclipse occurs when the Earth is directly between the sun and the moon. When in this position, the Earth blocks the sun's light rays from hitting the moon, leaving it in complete darkness. For an observer on Earth, the moon seems to fade because there is no light reflecting off the moon for us to see. As the Earth moves out from between the sun and the moon, the moon gradually becomes illuminated until the entire full moon is seen clearly once again.

What is a **solar eclipse**?

An eclipse of the sun occurs when the moon casts its shadow on the Earth. As the moon moves between the Earth and sun, darkness falls upon the section of the Earth where the moon blocks the light.

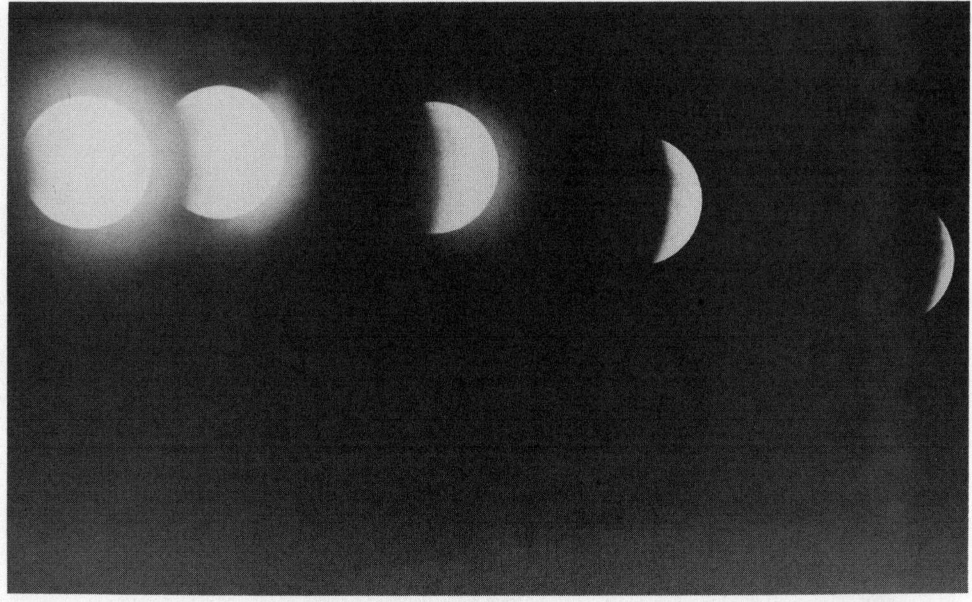

Five phases of a lunar eclipse are pictured in multiple exposures over the skyline of Toronto, Ontario, Canada, on August 16, 1989.

How **dark** can it get on Earth during a solar eclipse?

Up to a 90% decrease in brightness can occur during a total eclipse of the sun. The remaining 10% is from the bending of light around the moon and perhaps the refraction of light from the surrounding areas that are not affected as much. The complete eclipse of the sun, known as totality, usually lasts for an average of about 2.5 minutes, but has been known to last over 7 minutes.

What is the difference between the **umbra and penumbra** of a shadow?

A shadow has two distinct regions. The penumbra, or partial shadow, is a section where some light has entered, resulting in an area that is not completely bright, but not dark either. The umbra is the area of the shadow where all the light from the source has been blocked, preventing any light from falling on the surface.

During an eclipse, the outer regions affected by the eclipse experience the penumbra shadow, because some light from the sun has been able to

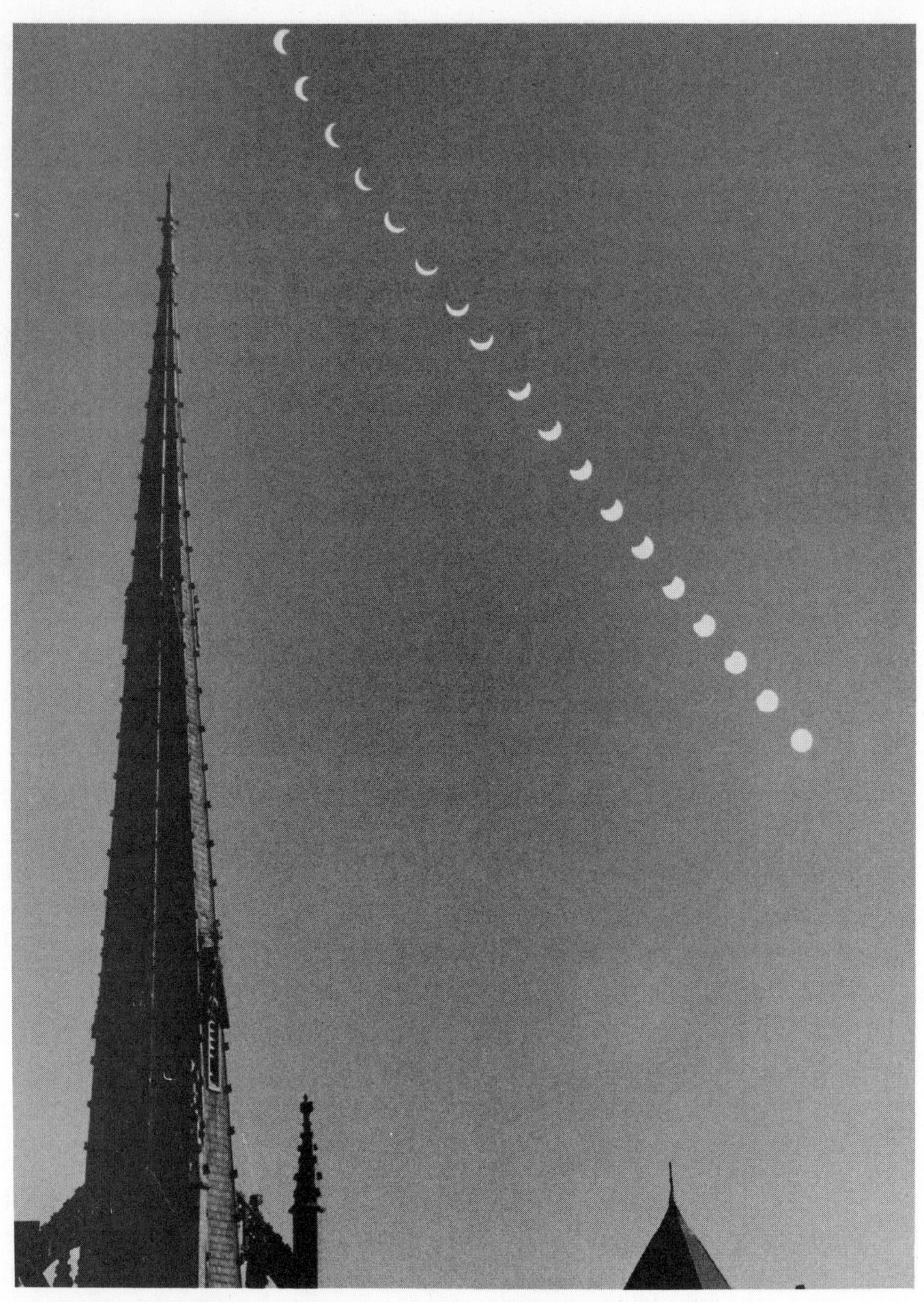

An eclipse of the sun captured over Detroit, Michigan, in a series of multiple exposures.

land on the surface. The area of total (or almost total) darkness is the umbra; no direct sunlight can reach this area, resulting in a complete eclipse.

Can umbras and penumbras be **adjusted**?

Umbras and penumbras exist where ever there are shadows. When an opaque object is close to the surface on which its shadow is cast, the shadow will be clear and distinct, because it has a large umbra section and small penumbra. However, if the object casting the shadow is placed farther away from the surface and closer to the light source, light hitting the edges of the blocked object will have a chance to bend around the object and shed some light on the fringes of the shadows. This produces a larger penumbra, while reducing the umbra section, resulting in a fuzzy or blurry shadow.

Since the **moon** has **monthly cycles**, why don't solar eclipses occur each month?

The moon has a monthly orbital cycle around the Earth, and if the moon were to keep the exact same orbital path every month, lunar and solar eclipses would happen a bit more regularly. However, the moon's position to the ecliptic (the Earth's plane of orbit around the sun) is tilted, and not on the same plane as the Earth's. Therefore, the moon and Earth are not often in alignment with respect to the sun—the position that creates an eclipse. The orbital motions of the Earth and moon do allow eclipses to be predicted well in advance, however.

How often do **solar and lunar eclipses** occur?

Solar eclipses (including partial eclipses) occur more frequently than lunar eclipses; there are usually two or three solar eclipses per year, while the average number of lunar eclipses per year is between one and two. Each occurrence of a solar eclipse can only be observed by a narrow section of the Earth, however, while a lunar eclipse can be seen over a much larger area.

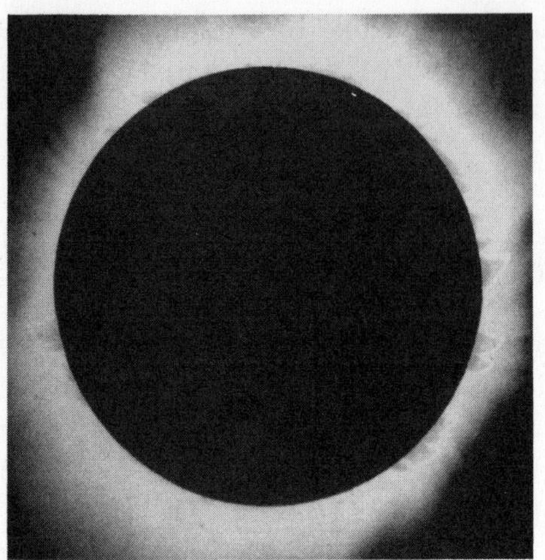

A total solar eclipse.

When a solar eclipses occurs, is the entire **Earth in the moon's shadow**?

To actually observe a total solar eclipse is quite rare, for the shadow of the moon only covers an area of about 300 kilometers in diameter. Therefore, a region lucky enough to experience a solar eclipse would most likely observe a partial one, with only part of the sun covered.

What is an **annular** solar eclipse?

Depending upon where the moon is in its elliptical orbit around the Earth, varying amounts of an eclipse will take place. If the moon is close enough to the Earth, a total eclipse could take place; however, if the moon is a bit further from the Earth, the moon might not be able to cover the sun. This would result in an "annular" solar eclipse, in which a bright ring of sunlight appears around the perimeter of the moon.

What are the **predicted occurrences** for **total solar eclipses** over the next few years?

The following chart shows the date and location of total solar eclipses that will take place over the next several years:

Date	Place
August 11, 1999	North Atlantic, Central Europe through India
June 21, 2001	Southern Atlantic, Africa
December 4, 2002	Southern Africa, Indian Ocean, Australia
November 23, 2003	Antarctica
April 8, 2005	Pacific Ocean, South America

The last total solar eclipse in the United States was seen on February 26, 1979. When will the United States see its next total solar eclipse?

Although the entire United States will not see the eclipse, it will sweep across a rather large portion of the country. The next total solar eclipse in the U.S. will be seen from Oregon to South Carolina on August 21, 2017. Mark your calendars!

What are the **predicted occurrences** for **lunar eclipses** for the next few years?

The following chart shows the date and the percentage of the moon that will be eclipsed by the Earth over the next several years:

Date	Percent eclipsed
July 28, 1999	42%
January 21, 2000	100%
July 16, 2000	100%

How did **ancient cultures** react to solar eclipses?

Many ancient cultures that worshiped the sun apparently felt that the sudden eclipse of the sun was a terrible occurrence. These events did not happen very often, but when solar eclipses did occur and partial darkness set in on the area, worshipers would gather and pray, often for days, to the "sun gods."

In the sixth century B.C., the Medes and Lydian armies were in a terrible battle when a solar eclipse occurred. The eclipse halted the battle and helped bring about a peace between the two forces; this disappearance of the sun was thought to be an omen for the two armies.

Why is it **dangerous** to look at a solar eclipse?

Looking at a solar eclipse is just as dangerous as looking at the sun under normal conditions (a very dangerous thing to do), because the sun light still bends around the moon. The sun is constantly emitting dangerous ultraviolet light that can cause great harm to your eyes—this is the same radiation that burns your skin. The retinas in your eyes are extremely sensitive to UV rays, and since there are no nerve endings on the retina to signal pain, the retina can easily burn and cause severe damage before a person realizes the effects.

Partial, usually temporary, damage can occur by looking at the sun even for a few fractions of a second. In fact, looking at the sun can leave an image of the sun on your retina for several minutes, so that everything you look at will have the sun on it as well!

The only way to safely look at the sun is to wear adequate and approved solar filters. Typical sunglasses will not do the trick. Dozens of people go to the hospital every time there is a solar eclipse because they feel they can safely look at the sun. Do not do this; instead make a pinhole camera (which will be covered later in this chapter) or get a special solar filter.

POLARIZATION OF LIGHT

What is **polarized light**?

Light is typically emitted in all directions and with different orientations; that is to say that a light wave may be oriented with the electric component of the electromagnetic wave vibrating up and down or vertically, while in other instances the light might vibrate horizontally or even diagonally. Regardless of the orientation, these light waves are not polarized. To be polarized, all the light waves must be oriented in the same direction. For example, vertically polarized light has its waves all aligned in up and down vibrations. Non-polarized light produces glare, which can be a distraction when driving, skiing, or taking pictures.

How can you check to make sure a pair of sunglasses is polarized?

Polarized lenses mean that they are only transparent to light that is aligned in one direction. Therefore, if a pair of sunglasses claim to be polarized, two pairs of the same sunglasses, when set up perpendicular to each other, should not allow any light to pass through the lenses. By aligning the lenses on top of each other and rotating them 90° from each other, the polarized gratings eliminate all orientations of light and therefore prove that the sunglasses are indeed polarized.

In addition, light from the sky is also partially polarized due to the scattering of light and gas molecules in the atmosphere, so if you put on a pair of polarized sunglasses and tilt your head so that your ear were near your shoulder, you should see a change in the intensity of the sky on a clear, sunny day. If you see no such change, then the sunglasses are not polarized.

How does light **become polarized**?

Sunlight and most other sources of light are emitted in an assortment of orientations and are non-polarized. To reduce glare, a polarizing filter is used to accept only one orientation. For example, if you want to accept only vertically polarized light, then a polarization filter with a vertical grating is used to allow all vertical vibrations to pass through, but prevent non-vertical light waves from entering.

Why are **polarized glasses** important?

Polarized glasses are useful when driving, sailing, skiing, or in any situation where unwanted glare is present. Glare is caused by light reflecting off a surface, such as water, a road, or snow. Navigating your way through some situations could be difficult without polarized sunglasses.

Take, for example, light reflecting off the surface of a lake. The light reflects in the same plane as the surface—horizontal, which can be extremely distracting to some people. In order to reduce or eliminate this glare from the water, vertically polarized sunglasses can be used. The vertical polarization filters permit only vertically oriented light waves to pass through the sunglasses, blocking the unwanted glare.

When looking through a pair of polarized sunglasses, why do the **rear windows** in cars appear to have **spots** on them?

The spots seen on rear windows when wearing polarized sunglasses are the stress marks of the safety glass. The spots, created during the manufacturing of the glass, act like polarizing filters and therefore block some of the light, creating small, circular, dark regions in the otherwise transparent glass.

Are the numbers in some **calculator and digital watch displays** polarized?

Electronic devices that display information through the use of LCD, or liquid crystal display, use rotating polarized filters to produce black segments that help form the numbers. When a segment needs to be displayed on the polarized screen, a small electric current is created to rotate another polarized section 90°, so no light passes through that section of the screen. The result is a black segment on the LCD display. If a black segment of the LCD is not to be displayed, the polarized filter turns another 90° to allow the light from the background to pass through, revealing a grayish or silver screen.

3-D

What is three-dimensional **(3-D)** vision?

Seeing in three dimensions, which is how a person with normal eyes sees, means that in addition to perceiving the dimensions of height and width (such as seen on a piece of paper, a poster, or a TV or movie screen), one can see the third dimension of depth. We see real objects in 3-D because we have two eyes (one is not just a spare) that see slightly different perspectives of the same view. The combination of these views, when interpreted and reconciled by our brain, gives us the ability to perceive depth, the third dimension.

If you close one eye, your ability to perceive depth is diminished. With only one eye, the world won't *look* very different to you (other than the loss of most of the right or left side of your field of vision, depending upon which eye you shut), but if you try to maneuver around, you'll experience difficulty in judging distances, and will feel a bit clumsy.

How do **3-D movies** work, since the movie is projected on a **two-dimensional screen**?

Although the movie is projected on a flat screen, a three-dimensional movie uses polarized glasses and two separate projectors to simulate the natural separation of eyesight. When a three-dimensional movie is filmed, two cameras film the movie from slightly different positions. When the film is projected on the screen, each projector uses a separate polarizing filter. The left projector might use a horizontally polarized filter, while the right projector uses a vertically polarized filter. The observer also wears 3-D glasses, which are polarized as well. Therefore, the left eye would see only the image produced from the horizontally polarized image of the left projector, while the right eye would only see the image produced by the vertically polarized right projector. This arrangement simulates the different perspectives that each eye sees when looking at a real-life 3-D scene, allowing the brain to interpret the difference as depth (the third dimension).

There are other methods of achieving simulated 3-D viewing, such as the older method using color filters rather than polarized filters, or a newer, more expensive method in which you wear goggles that react in synchronization with alternating (rather than superimposed) images on the screen, but all the methods depend on presenting a slightly different image to each eye.

What would happen if a person did not wear the **3-D glasses** when watching a movie?

Anyone can still watch a three-dimensional movie without the glasses; however, the drawback is that the images might seem blurred at times, depending upon how dramatic a 3-D effect the film is producing. And just for fun, if you put the glasses on backwards—that is, looking through them from the opposite direction—the depths will be reversed; people in a scene will appear to be farther away than the background of the scene.

COLOR

What color is **white light**?

White light is the combination of all the colors in the visible light spectrum. When separated from each other, the different frequencies create different colors: the lowest-frequency light is the color red, and increasing frequencies result in orange, yellow, green, blue, indigo, and finally, the highest-frequency visible color, violet.

How are objects **seen**?

In order to see an object, it must be emitting or reflecting light. We can see stars, lightning, and light bulbs because they are emitting or giving off light. And we depend on the light emitted from these objects in order to see objects that don't emit light—we see those objects because they reflect light. A blade of grass, for example, does not emit light; however, we see the blade of grass because it reflects existing light, specifically green light.

Why do we see **specific colors**?

When we "see" colors, we are actually seeing the effect of light shining on an object. When white light shines on an object it may be reflected,

What is color blindness?

Some people are unable to see some colors due to an inherited condition known as color blindness. John Dalton, a British chemist and physicist, was the first person to discover color blindness in 1794. He was color blind himself, and could not distinguish red from green. Many color-blind people do not realize that they cannot distinguish colors accurately. This is potentially dangerous, particularly if they cannot distinguish between the colors of traffic lights or other safety signals. Those people who perceive red as green and green as red are known, appropriately, as "red-green color blind." Other color-blind people are only able to see black, gray, and white. It is estimated that 7% of men and 1% of women are born color blind.

absorbed, or transmitted (allowed to pass through). Glass transmits most of the light that it comes in contact with, and thus appears colorless. Snow reflects all of the light, and so appears white. A black cloth absorbs all light, and so it appears black. A green blade of grass reflects green light better than it reflects other colors. Most objects appear colored because their chemical structure absorbs certain wavelengths of light and reflects others.

Who discovered that **white light** could be separated into the colors of the **rainbow**?

Chandeliers were quite popular in the seventeenth century, after great improvements were made in the manufacturing of glass. Newton, who was intrigued by the colors that were produced by those chandeliers, decided to examine how a piece of glass—actually a prism—could create such a spectrum of colors. He set up an experiment in which he darkened his room in Cambridge, England, except for a small hole that he made in a shutter. Newton positioned the prism so that the white sunlight could pass through, creating a beautiful spectrum of colors on the opposite wall of the room.

Perhaps a larger breakthrough for Newton was the fact that he could reverse the procedure and form white light from the spectrum of colors. He accomplished this by placing a lens in the middle of the spectrum to keep the colors of light parallel to each other and placed another prism in the path of the colors. Sure enough, a beam of white light emerged out of the second prism.

If white is the combination of the colors of the rainbow, what is **black**?

Black, the exact opposite of white light, is the absence of light or the absorption of all light. A black piece of paper appears black because all the light is being absorbed into the paper—none is reflected back out.

What are the **primary colors** in emitted light?

When mixing paint, any first-grader knows that red, yellow, and blue—the primary colors—can be combined to make all the other colors. When mixing light (or "additive color mixing"), however, there is a different schema. The three primary colors in light are blue, green, and red. The combination of these primary colors results in variations of other colors, and when all three colors are combined with equal intensity, a color very close to pure white is formed. The primary colors are different for light because colorants (such as pigments, inks, and dyes) *reflect and absorb* light, rather than emit it.

What are the **secondary colors**?

When any two of the three primary colors are mixed, secondary colors are formed; they are called secondary because they are by-products of the primary red, green, and blue colors. Like primary colors, secondary colored light is different than the secondary colors that are formed when mixing colorants. Red light mixed with green creates yellow light. Red and blue produces magenta. Finally, cyan is formed when blue light and green light are added together.

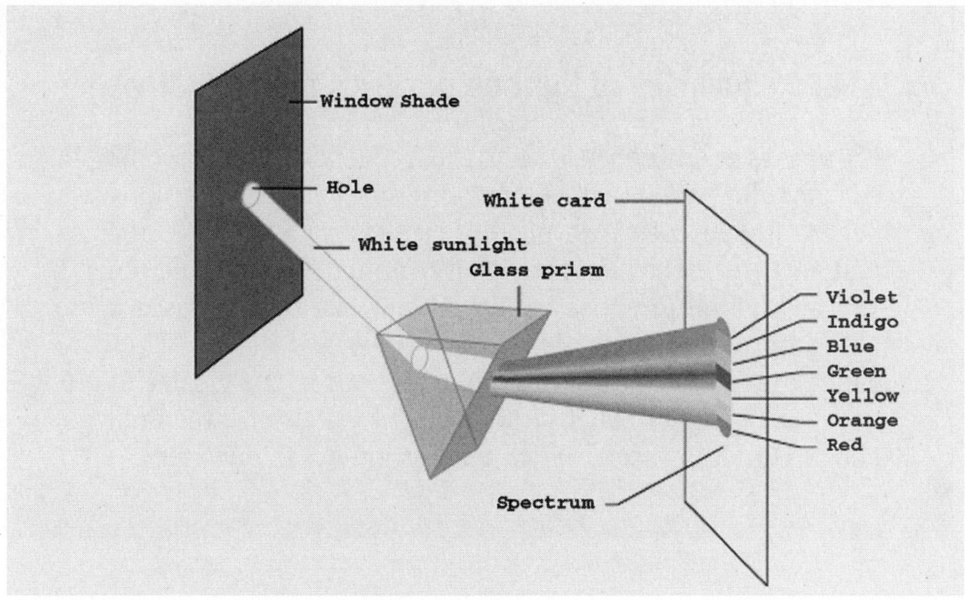

A diagram of Newton's 1666 spectrum experiment.

What are **complementary colors**?

Complementary colors are pairs of one secondary and one primary color that, when mixed, form what is close to white light. For example, yellow and blue light are complementary because when combined, they form white light, as will magenta and green, and cyan and red.

What is **subtractive color mixing**?

As opposed to the mixing of light ("additive color mixing"), subtractive color mixing occurs only when combining dyes, pigments, or other objects that absorb and reflect light. Examples of things that use subtractive color mixing are paints and inks. When colors are mixed in paints and inks, black is formed, as opposed to the white that results when colored lights are added to each other.

What are **primary pigments**?

The primary pigments are the primary colors for colorants such as inks, paints, and dyes. These are the same colors as the secondary colors

What frequencies of light do our eyes most readily see?

Our eyes are most sensitive to the frequencies corresponding to the yellow and green colors of the spectrum. Flashy signs and modern fire engines are typically painted a yellowish-green color to attract a great deal of attention from us. Even simple objects such as highlighters, used to emphasize words or phrases while taking notes, are typically bright yellow and green. When we glance over something or see an object out of the corner of our eyes, we are more likely to notice bright yellowish-green objects than red or blue objects, which are less intense to our eyes.

obtained when mixing light (additive color mixing). Primary pigments or dyes are magenta, which reflects blue and red light but absorbs green; cyan, which reflects blue and green but absorbs red; and yellow, which reflects red and green but absorbs blue.

What are the **secondary colors** in subtractive color mixing?

The secondary colors for dyes and pigments are the same as the primary colors in additive color mixing. However, the red, green, and blue reflect their own color while absorbing the other two colors. For example, red would reflect red but absorb the green and blue light.

Why do most color ink-jet printers use four colors to print, instead of the **three primary colors for subtractive color mixing**?

It would seem that a color ink jet printer mixing the three primary pigments of yellow, magenta, and cyan should be able to produce all the other colors, including black. However, when all three primary colors are combined, the mix looks more like a muddy brown color than black.

Why is the sky blue?

Thousands of parents are asked this question by their inquisitive children, and many have no idea as to why the sky is indeed blue. The answer to this age-old question is one word: scattering. When white light encounters the oxygen and nitrogen atoms in the Earth's atmosphere, the high-frequency section of the white light strikes the orbiting electrons of the nitrogen and oxygen, which causes the high-frequency light to be scattered in all directions. The high-frequency light scattered by the molecules in the air are the violet, indigo, and blue colors of white light. These are the colors reflected by the atoms in the atmosphere, and therefore are the colors that we see in the sky.

Although these are the primary colors of which other shades can be created, they do not represent all the colors of the spectrum needed to form black. Therefore, most of today's color ink jet printers have a cartridge with yellow, cyan, and magenta ink, and another separate ink cartridge of just black.

What is **colorimetry**?

Because the perception of color is mostly a neurophysiological function between the eyes and the brain, it can vary slightly from person to person. Scientists, artists, advertisers, and printers need an objective method of specifying color as it relates to the frequency of light. This technique for measuring the particular frequencies of light is known as colorimetry.

What is the difference between **hue and saturation**?

Hue is the particular frequency of a specific color of light. Saturation is the extent to which other frequencies of light are present in a particular

Why are snow and clouds white?

Snow and clouds consist of different sized water droplets. Small droplets scatter high-frequency light, and large droplets scatter lower-frequency light. Together, the water droplets absorb little light energy, but scatter all the colors of the spectrum to create a white reflection.

color. A color with pure hue and full saturation is rarely seen outside the artificial setting of a laboratory.

If the high-frequency light is scattered, why do we only see a blue sky and not a **blue, indigo, and violet sky**?

Our eyes are most sensitive to colors in the mid-section of the color spectrum. Because blue is closer to the mid-section of the color spectrum, it is perceived by our eyes more readily than the indigo and violet. Even though all three colors are scattered by the molecules and particles in the air, humans see a predominantly blue sky.

On humid summer days, why does the sky take on a **white or grayish** appearance?

When high amounts of humidity are in the air, water molecules are more prevalent than on a cool, dry day. Water molecules, which have two hydrogen and one oxygen atom, are larger than oxygen and nitrogen found in the air, and the size of a molecule plays a significant role in what frequencies of light are scattered. When white light encounters a larger molecule or dust particle, lower-frequency light will be scattered, whereas if a smaller molecule is struck by white light, higher-frequency light will be scattered.

Because humid days have more water droplets in the atmosphere, the lower frequencies of red, orange, yellow, and green are scattered. However, in areas where white light strikes small water molecules or only oxygen or nitrogen molecules, blue, indigo and violet light is scattered. The combination of the different frequencies of light results in a white or, if less intense, a gray-colored sky.

If blue light is scattered by the small molecules in the Earth's atmosphere, wouldn't all the **blue light** be used up by the time it reached the ground?

Only a small percentage of blue light is scattered in the middle of the day. Since the atmosphere is relatively thin, there is plenty of blue light left when the sunlight strikes the surface of the Earth.

Why are **sunrises and sunsets** often orange or red?

During the evening and early morning, when the sun is lower in the horizon, the light that the sun emits has to travel farther and through more of the atmosphere to reach us than it does during mid-day, when the sun is closer to us. Since the distance through the atmosphere is much larger for sunlight in the morning and evening than during mid-day, the blue, indigo, and violet frequencies get scattered out and used up in the evening and morning. When the light finally reaches us, the only frequencies left are those low-end frequencies of red, orange, yellow, and a little bit of green (some green has also been scattered). The exhaustive scattering of the blue light is responsible for the beautiful red, orange, and yellow sunrises and sunsets that we all love to watch.

When a sunset is **red or orange**, why is the sky directly above us still **blue**?

Although the sunlight traveling from the sun and through the long distance of the Earth's atmosphere has scattered out the blue light, the sky above us is still blue. Not all the sunlight scatters through the thick section of the Earth's atmosphere—instead, some of it skims the section of the atmosphere directly above us. Since only a small amount of sunlight has hit this section, there is plenty of high-frequency light to be scattered, so the sky above appears blue, even though the sky by the setting sun is red, orange, and yellow.

Why is the **ocean blue**?

There are two major reasons why the ocean and most bodies of water appear blue. The first can be observed by looking at the water on a cloudy day and then on a sunny day. There is a rather large difference in how blue the water appears to be on the two different days, because the water seems to act as a mirror for the sky. So on a sunny day with a blue sky, the water will have a richer blue color than on a cloudy day.

The second reason why bodies of water have a blue appearance is that water tends to reflect or scatter high-frequency light more than the low-frequency colors. In fact, water absorbs low-frequency light such as infrared, which helps the temperature to increase, and absorbs red and a little bit of orange light. The result is the reflection of yellow, green, blue, indigo, and violet. The overwhelming amounts of reflected, high-frequency light results in a crisp blue-colored body of water.

Some bodies of water may take on a more greenish or at times a brownish or black color. Usually this is due to other elements in the water such as algae, mud, sand, minerals, and increasing instances of pollution in the water. Still, in the overwhelming majority of cases, water looks blue.

RAINBOWS

How do **rainbows** occur?

A rainbow is a spectrum of light formed when sunlight refracts into, reflects, and then refracts once again out of water droplets. Upon entering a water droplet, the white light is spread apart into its individual frequencies, just as in a prism. The light inside the droplet then reflects against the back of the water droplet and spreads apart even more upon exiting the water droplet. The separation of the light frequencies, along with a large number of water droplets exposed to sunlight, creates a circular-shaped rainbow.

What **conditions** must be met in order to see a rainbow?

There are only two main conditions for witnessing a rainbow. The first is that the observer must be between the sun and the water droplets. The water droplets can either be rain, mist from a waterfall, or the spray of a garden hose. The second condition is that the angle between the sun, the water droplets, and the observer's eyes must be between 40 and 42°. Therefore, in order to see a rainbow from rain, the rainbow needs to occur in the morning or afternoon when the sun and observer are between 40 and 42° of each other.

What is the **order of colors** in a rainbow?

The order of colors in a rainbow goes from low frequency on the outer arc to higher-frequency light on the inside of the arc. The full order from outer to inner is: red, orange, yellow, green, blue, indigo, and violet.

Who was the **first person** to determine how rainbows occurred?

Newton was not the first person to understand the optical characteristics of a rainbow. In fact, it was a German monk in the early fourteenth century who first discovered that light refracted and reflected inside a water droplet. To demonstrate his hypothesis, the monk filled a sphere with water, sent a ray of sunlight through the sphere, and observed the separation of the white light into colors along with the reflection on the back of the water droplet.

What is a **supernumerary rainbow**?

A supernumerary rainbow is another name for double rainbow. The secondary rainbow has its color spectrum reversed, is outside of the original rainbow, and is significantly dimmer than the primary rainbow. The

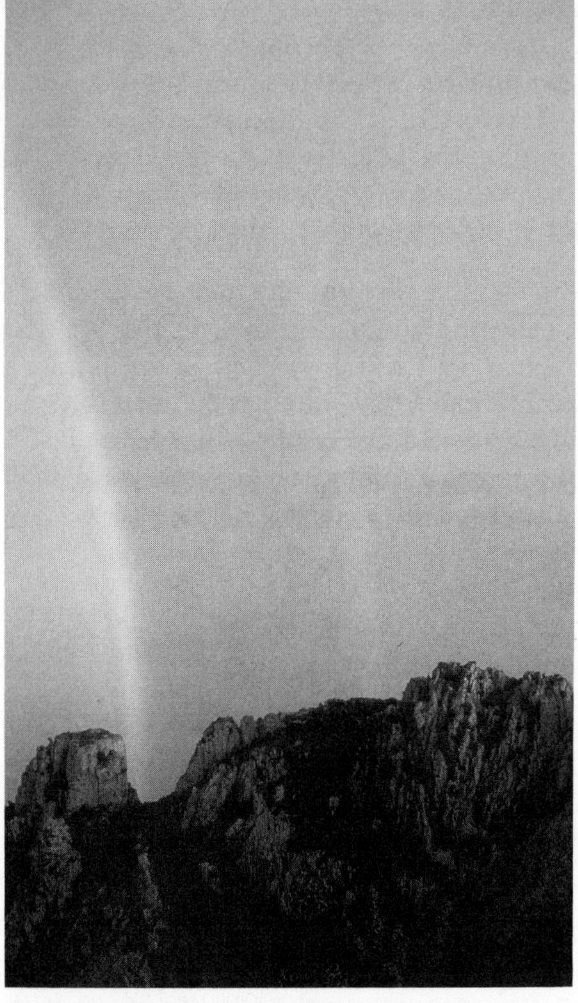

A supernumerary (double) rainbow.

reason why a supernumerary rainbow occurs is because an additional reflection of the light takes place inside the water droplets. Instead of reflecting once in the water droplet, the light reflects twice inside the water, creating a dimmer mirror image of the primary rainbow.

Does everyone see the **same rainbow**?

Since a rainbow is dependent on the position of the sun, the water droplets, and the observer, everyone watching a rainbow is actually seeing a slightly different rainbow.

Is there such a thing as a completely circular rainbow?

All rainbows would be completely round except that the ground gets in the way of completing the circle. However, if viewed from a high altitude, such as an airplane, circular rainbows can been seen when the angle between the sun, the water droplets, and the plane is between the 40 and 42°. In this case, the rainbow is horizontal, meaning that it is parallel to the ground and therefore not blocked by the ground. This is quite a sight!

OPTICS

What is **optics**?

Optics is an area of study within physics that deals with the behavior and movement of light. Optics does not deal solely with visible light, but with the other sections of the electromagnetic spectrum as well, including microwaves, infrared, visible, ultraviolet, and X-rays. Optics in its own right has two major subdivisions, physical and geometrical optics.

What is the difference between **physical and geometrical optics**?

Geometrical optics deals specifically with the path that light takes when it encounters mirrors and lenses. Geometrical optics ignores the wave theory of light, and uses ray diagrams to trace and understand the path that light takes when it reflects and refracts in different media.

Physical optics, on the other hand, is the subdivision of optics that studies the more complicated behavior of light found in polarization, diffraction, interference, and the spectral analysis of light waves.

REFLECTION

What is **reflection**?

Reflection of light occurs when light "bounces" off of a surface, such as a mirror. The amount of reflection depends upon the nature of the surface of the object that the light is striking. A surface that absorbs the light energy will not reflect back as much light as a surface that does not absorb light. Secondly, a surface that is rough and irregular will cause the reflected light to scatter, making it extremely difficult to see a clear image.

Polished, smooth surfaces that do not absorb light are best at reflection; examples of reflective materials are shiny metals, whereas non-reflective materials are dull metals, wood, and stone.

MIRRORS

How were the **first mirrors** made?

People have seen their reflections in water for hundreds of centuries, but some of the earliest signs of human-made brass and bronze mirrors have been mentioned in the Bible and in ancient Egyptian, Greek, and Roman literature. The earliest glass mirrors, backed with shiny metal, appeared in Italy during the fourteenth century. The original process for creating a glass mirror was to coat one side of glass with mercury and polished tinfoil.

The method similar to what is used today for silvering a mirror was discovered by German chemist Justus von Liebig in 1835. His process consisted of pouring ammonia and silver compound onto the glass surface. When formaldehyde was added to the metal, it produced a shiny silver surface that reflected the light. Today, mirrors are made in different shapes, angles, and reflective surfaces to achieve the desired effects.

Videotaping done via a one-way mirror.

Why can't you always **see yourself** in a mirror?

Disregarding the possibility that you are a vampire, the answer has to do with angles. The law of reflection states that the angle of light incident on a mirror must be equal to the angle of the reflected light. If you are standing directly in front of a mirror, your angle of incidence (that is, the angle between the direction the incoming light is traveling, and a line that is perpendicular to the surface) might be zero and as a result, reflects directly back at an angle of 0°. However, if the angle is significant, such as the angle between your eyes and a rear-view mirror in a car, it may be so large that the reflected angle does not come back to your eye and instead you see out the back of the car.

How do **one-way mirrors,** the ones used in interrogation rooms, work?

A one-way mirror acts as a mirror from one side, but as a window from its opposite side, effectively "disguising" a window as a mirror to allow secret surveillance. There are two conditions that must be met in order for a one-way mirror to perform properly. First, the interrogation room must be brighter than the observing room behind the mirror. It is more difficult for a person being interrogated to look into a dark room from a bright room. Secondly, the silver backing of the mirror must have only half the normal amount of silver. This allows some light to be reflected back into the interrogation room, and some light to pass through into the observation room. It is imperative that the observation room remain dark, because if a lamp were turned on, some of that light would pass through into the interrogation room as well.

On the front of many **ambulances,** why is the word "ambulance" printed **backwards**?

The word "ambulance" is printed backwards so that, when viewed in a mirror—specifically, the rear-view mirror of a car—it will appear correctly (that is, forwards), ensuring that the driver can respond appropriately in the quickest possible manner.

How do day/night **rear-view mirrors** function in vehicles?

When drivers traveling at night encounter a bright light from the vehicle behind them shining in their eyes, many will flip a tab on the underside of the rear-view mirror to deflect the light up toward the ceiling of the car. The silvered surface of the mirror reflects approximately 85–90% of the incident light on the mirror, which is now directed toward the ceiling. The remaining 10–15% of the light is reflected by the front of the glass on the mirror. That glass is slightly angled downward to allow the remaining light to be reflected into the eyes of the driver. Since the amount of light is greatly reduced, the light should no longer bother the driver.

What is a **virtual image**?

A virtual image is an image that appears to be located behind a mirror or lens. When you look at yourself directly in a mirror, your image appears to located on the other side of the mirror—this is known as a virtual image.

Why do side-view mirrors in vehicles state: "Objects viewed in mirror may be closer than they appear"?

This statement, seen on most side-view mirrors, is a very important safety message—the message warns the driver that the mirror is deceiving. Why would an automobile manufacturer put a deceiving mirror on a car? A flat, plane mirror would only show the driver a small, narrow section of the road behind the car; however, if a convex mirror is used, the driver can not only see behind the car, but to the side as well, reducing his or her blind spot. In the process, however, convex mirrors make objects appear smaller and therefore farther away, so the message is there to serve as a reminder that the image is not exactly as it appears.

Virtual images do not have light originating from them, but they appear as if they do. In addition, a virtual image cannot be focused on a screen.

What is a **real image**?

A real image is an image that has light originating from it and that can be projected on a wall after it has passed through a lens or reflected off a mirror. Real images are typically more useful than virtual images simply because they can be made larger than the original object, and can be focused and projected on a screen or wall.

What is a **concave mirror** and what is it used for?

A concave mirror is curved inward so that the incident rays of light are reflected and focused in on one point, called the focus. Concave mirrors are typically used to focus wave energy, whether it is a microwave signal to a receiver, or visible light to an observer. When looking into a concave bathroom mirror, the image of a person within the focal point is upright, while the image further away from the focal point appears upside down.

What is a **convex mirror** and what is it used for?

A convex mirror is the exact opposite of a concave mirror. The convex mirror is curved outward, allowing the reflected light not to focus in on one point but to spread out. Convex mirrors are used for security purposes in stores because they broaden the reflected field of vision, allowing clerks to see a large section of the store. Although the objects viewed in convex mirrors are smaller than they are in real life, the mirrors help to see a wide area.

REFRACTION

What is **refraction**?

Light can be redirected in three ways: reflection, diffraction, or refraction. Reflection occurs when light bounces off of a surface, a mirror, for example. The second method of redirection, diffraction, is the deviation from a straight path that occurs when a wave such as light or sound passes around an obstacle or through an opening. The third method, refraction, refers to the bending of light as it goes from one medium to another. Eyeglass lenses refract light so that the wearer's eyes can focus it properly. Sunlight refracts when it encounters the medium of the Earth's atmosphere, and when it goes through water. An image can look quite different after viewing it through a refracting medium.

When a light wave encounters a **different medium,** can the refraction of light be determined?

The extent to which a beam of light bends when it hits a different medium depends on the physical characteristics of that particular medium and the angle at which the light enters the new medium.

Refraction in water causes the swizzle stick to appear broken at the water line.

All material has an index of refraction. It can be thought of as a measure of how quickly light moves through the material. The index of refraction for a given medium is the speed of light in a vacuum divided by the speed of light in that medium. Therefore, a vacuum, with no impediment to light propagation, has a refractive index of 1, whereas glass has a higher value, typically around 1.5. The higher the index of refraction, the more slowly light travels through the medium.

According to the Law of Reflection, light that hits a surface at a given angle is reflected off at that same angle. This principle makes it possible to see around a corner using a mirror held at 45°.

Snell's Law of Refraction, named after Dutch physicist Willebrod Snell, tells us how light behaves at a boundary between two different kinds of material. According to Snell's Law, when light hits the boundary, or interface, between two materials, it is bent from its original path (refraction). Consider the interface between two materials where the refractive index of the top material is lower than that of the bottom material. According to Snell's Law, the light will be bent from its original path to a smaller angle, closer to the line perpendicular to the surface of the second material. As the incoming, or incident angle increases, so does the refracted angle.

What is the **index of refraction** for light traveling through different media?

The following are some sample indexes of refraction for various materials. The index of refraction, represented by "n," is the ratio of the speed of light in a vacuum divided by the speed of light in the material. The larger the index of refraction, the greater the bending that takes place.

Medium	Index of Refraction (n)
Vacuum	1.00
Air	1.0003 (usually rounded to 1.0)
Water	1.33
Crown Glass	1.52
Quartz	1.54
Flint Glass	1.61
Diamond	2.42

By knowing the above indexes of refraction and the angle that the light approaches the new medium, a person can determine exactly how the light will refract when it encounters the medium.

Are the **stars** really where they appear to be?

First of all, the light from stars has traveled for millions of light-years, so there is a good chance that the stars we see no longer exist. However, this question is referring to the position of the stars. The light from

Why does a person standing in a pool of water often appear short and stocky?

The portion of a person's body that is above water does not appear out of proportion because the light entering your eye is not going through a different medium and refracting. However, the portion that is underwater—the person's legs—appears to be short because the light reflecting off their legs is traveling through water and then into the air. Due to this change in medium, refraction occurs. Since the index of refraction for water is larger than the index of refraction for air, the legs of the person appear compressed and stocky.

stars refracts slightly as it enters our atmosphere on Earth. Therefore, the true position of the stars is probably a bit off from where we think they are. The same thing goes for the sun and moon, especially when they are low on the horizon; the light from the sun and moon is refracted slightly, resulting in a distorted view of the moon and sun when in these positions.

LENSES

When did **lenses** first appear?

The ancient Greeks and Romans did not have lenses in the true sense of the word, but did manage to refract light with water-filled glass jars and spheres. It was the Arabs who realized that lenses could be made to magnify objects. They used magnifying lenses a great deal for reading, and eventually someone in thirteenth-century Europe placed two magnifying lenses in a frame and made the first pair of spectacles or glasses. The lenses were made of beryl, a precious stone that was transparent to light and easy to shape into a magnifying lens. Today, lenses are made using transparent and lightweight plastics that are cheaper and more durable than traditional glass lenses, which are heavy and fragile.

What is the **focal length** on a lens?

The focal length of a lens is the distance from the center of the lens to where the rays of light transmitted through the lens would meet. A very round lens would have a short focal length, while a more flat lens would have a very long focal length.

What is a **diverging lens**?

A diverging lens has at least one concave side; the other is usually flat. The shape of the lens causes the entering light rays to emerge with a greater divergence (that is, a greater spread in the direction the rays are traveling) than they entered with. This allows a diverging lens to create a larger image when projected on a screen.

What is a **converging lens**?

A converging lens has at least one convex side. Its shape causes the entering light rays to converge, this is, the direction the light rays are traveling would eventually intersect if they could travel a sufficient distance. This creates a small, upright image before the focal point, but after the focal point, the image inverts and can be larger than the object when projected on a screen.

How does a **pinhole camera** produce an image?

A pinhole camera can be a shoe box with a small "pinhole" in one side of the box and a screen on the other side of the box. When light enters the hole in the box, the hole acts as a converging lens. It brings the rays of light together to a focal point (the pinhole); beyond the focal point, it inverts and projects the image on the screen.

Eclipse as seen through a pinhole camera.

Pinhole cameras are easy to make, and are often used during solar eclipses because it is very dangerous to look directly at the sun (during an eclipse or otherwise). With the sun at your back, point the hole up toward the sun and view the image of the moon passing in front of the sun on the screen.

What is a **mirage**?

Mirages typically occur on hot summer days when surfaces such as sand, concrete, or asphalt are warm. Mirages look like pools of water on

A mirage in the desert.

the ground, along with an upside-down image of a building, vehicle, or tree in the distance. As one approaches a mirage, the puddle of water and the reflection seem to disappear.

A mirage occurs because of a temperature difference between the air directly above the surface, which is hot, and the cooler air a few meters above the surface. The difference in the index of refraction of the two different air temperatures causes the light from an object to bend up toward the observer. As a result, the object is right side up while the refracted image is inverted and underneath the original object. The illusion of water is also a refracted image, the image of the sky. Mirages can only occur on hot surfaces and objects that are at relatively small angles in relation to the observer. Therefore, a person cannot see a mirage for an object that is just a few meters away. A mirage is not a hallucination, but instead a true and well-documented optical phenomenon.

What is **total internal reflection** and **critical angle**?

Total internal reflection occurs when a ray of light, attempting to exit one medium and enter another, is reflected back into its original medium. The light cannot escape because it is striking the boundary between the media at too large an angle, called the critical angle. When the ray of light is reflected back into instead of refracted out of the medium, it has reached its critical angle and is totally internally reflected.

Why do diamonds **sparkle** so much?

One of the keys to finding a good diamond is making sure it has a good cut. The sides of the diamond need to be at specific angles so that when light enters the diamond, it is internally reflected instead of being refracted out of the diamond. The critical angle for a diamond is about 25°; since the critical angle of the diamond is so small, this ensures that most of the light that has entered the diamond will emerge not from the sides but from the top of the diamond, resulting in a sparkle of shimmering light.

If you open your eyes **underwater,** can you see out of the water?

An example of total internal reflection can be witnessed when underwater. If a diver looks straight up out of the water, he or she will see the sky and any other visible surroundings directly above the water. However, if the diver looks out of the water at an angle of 48° or more from the vertical, the diver will not see out of the water, but instead will see a reflection from the bottom of the lake. The next time you are in a pool, try looking up out of the water and you will see a point on the surface where you no longer can see out of the pool, for the light has reached its critical angle.

Fiber Optics

How do fiber optic cables use **internal reflection** to transmit information?

Strands of glass fiber, commonly known as fiber optics, use the principle of total internal reflection to transmit information at the speed of light. Modern lasers send light into the end of a strand of fiber to send infor-

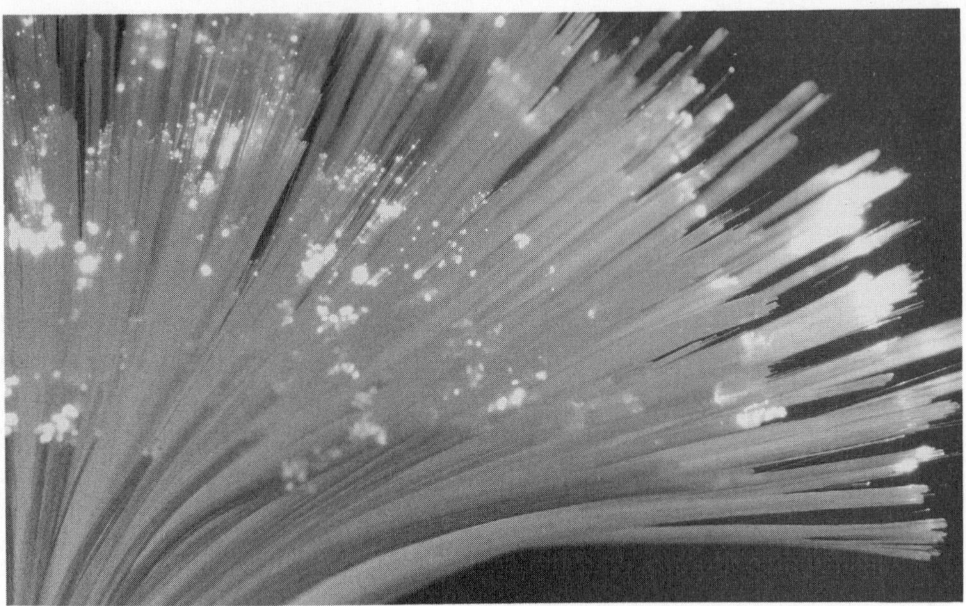

Optical fibers.

mation to the other end. When the laser light strikes the outside wall of the glass fiber, the light, instead of refracting out of the glass, reflects back into the cable, moving its way down the length of the fiber. The cladding of the optic cable has as high an index of refraction as possible, so the critical angle is as small as possible. Only light that strikes the cladding boundary inside the critical angle will refract outside the cable.

The low critical angle of glass against an outer cladding permits the light to travel great distances without significant damping. This new technology of transmitting information has and will continue to improve the field of telecommunications dramatically.

How did fiber optics **originate**?

The idea that light could travel through glass strands originated as far back as the 1840s, when physicists Collodon and Babinet demonstrated that light could travel through bending water jets in fountains. The first person to display an image through a bundle of optical fibers was a medical student in Germany by the name of Lamm, who, in 1930, used a bundle of fiber to project the image of a light bulb. In his research,

Lamm ultimately used optical fibers to observe and probe areas of the human body without making large incisions. From that point on, serious research in optical fibers ensued, later on exploding with the development of the laser.

Where are fiber optics used **today**?

The transmission of light information through optical fibers has had a huge impact on the technological world. The medical field has benefited greatly from the use of fiber optic bundles, which enable the viewing of areas of the body that would otherwise be inaccessible. A laser beam travelling through a fiber can also be sent into the body without performing major "open" surgery.

Communications is probably the field that is benefitting the most from the advent of fiber optic technology. Systems within area computer networks are using fiber optic cables to speed transmission times for applications and files. Light transmitted through the optic cables can go on for hundreds of kilometers before a boost is needed in the signal, a significant improvement over the conventional system of electrical transmission of information.

What are some major benefits of using light instead of electricity to **transmit information**?

There are many fundamental benefits to using light to transmit information. First, there is little or no heat generated when transmitting light information, as there is with electricity. Electrical circuits heat up over time and need to be cooled. Secondly, light travels much faster than conventional electrical signals. Another benefit to using light is that electrical interference does not distort the signal, as it does in electrical information transfer. Fiber optic wires are more amenable to bending; copper wiring increases its resistance if bent. In addition, fiber itself is much cheaper than copper wiring. Finally, fiber can carry much more information. One strand of fiber, along with the help of modulating lasers, could transmit the telephone calls and television stations for an entire town.

DIFFRACTION

What is the **diffraction of light**?

The diffraction of light occurs when light bends around the boundaries of an object attempting to create a shadow. When light is sent through an opening that is wider than the wavelength of the light, a clear and distinct shadow is formed. However, if the light needs to pass through a very narrow opening, it bends around the edges of the opening and a fuzzy shadow is created. Diffraction occurs with all types of waves, but can be most often and easily seen with light.

OPTICAL INSTRUMENTS

EYESIGHT

How does the **human eye** see?

The eye does not "see"—it only receives stimuli that are transmitted to the brain for translation and synthesis. In order to receive images, our eye has a lens to focus the image, an iris to regulate the amount of light entering our eye, and a screen called the retina.

In order to focus in on objects that we want to see, our eye changes the focal length and shape of our lens and cornea by contracting or relaxing the ciliary muscle around the eye. Once in focus, the image is inverted by the double convex lens and focused on the retina. The retina, packed with millions of light-sensitive cones and rods, sends the electrical impulses to the optic nerve and ultimately to the brain, where the image is translated and inverted to form a right-side up image.

What is the difference between **cones and rods**?

Cones are cone-shaped nerve cells on the retina that can distinguish or pick up fine details in images. They are located predominantly around

Why do soap bubbles and gasoline spills create different color reflections?

Iridescence is the reason that different colors are produced when light hits a thin film such as a bubble or gasoline. Iridescence is caused by the interference of light waves resulting from multiple reflections of light off of surfaces of varying thickness. A soap bubble displays an iridescent pattern because light reflects off the top and bottom surfaces of the soap bubbles. As the thickness of the layer varies, the color seen will vary.

Gasoline spills are easily seen on wet roads; this is not because people spill more gas when it has rained, but is instead due to the iridescent patterns that result from light reflecting off of the top of the gasoline, the underlying surface of the gasoline, and the water surface. The resulting pattern appears as the colors of the visible light spectrum in the thin film of gasoline.

the center of the retina. This area of the retina—called the fovea—picks up images in sharp detail. The cones also enable color vision.

As the distance grows from the fovea, rod-shaped nerve cells replace the cones. The rods are responsible for a general image over a large area, but not fine details. This explains why we look at objects straight on when examining something carefully. The image will be focused around the fovea, where the majority of cones pick up the fine details of the image. In addition, the rods are used for night vision.

What is the **shape of a lens** when focusing on objects far away and objects up close?

The ciliary muscle, responsible for changing the shape of the lens, adjusts its tension to focus on different distances. When focusing on objects far away, the lens needs a large focal length, so the muscle is relaxed in order to make the lens relatively flat. However, when an object is closer to the eye, a shorter focal length is needed. The ciliary muscle contracts, reducing the focal length of the lens by making it more spherical. The process of adjusting the shape of the lens to focus in on objects is called "accommodation."

When swimming underwater, why is vision blurred when you open your eyes, but clear when wearing **swimming goggles**?

Although the eye's lens changes shape to focus images on the retina, most of the refraction of light takes place during light's transition from air to the cornea. When water is substituted for air, that changes the amount that light is refracted in the cornea, producing a blurred image on the retina. However, if even a small amount of air is in front of the cornea, the refractive properties will return to normal.

How **close** can an object be before it appears **blurry**?

There is a limit as to how close an object can be to the eye before the lens can no longer adjust its focus. Up to about thirty years of age, the closest an object can be focused is approximately 10 to 20 centimeters. As one grows older, the lens tends to stiffen and it becomes more difficult for the person to focus on close objects. In fact, by the time a person reaches the age of seventy, their eyes cannot focus on objects within several meters of their eyes. As a result, most aging adults need reading glasses to focus on close objects.

What is **nearsightedness** and what can be done to correct it?

Nearsighted vision means that a person can only see objects that are relatively near the eye. When a person is nearsighted, it means that the relaxed, relatively flat lens in the eye has a focal length that falls just short of the retina. Depending upon the extent to which a person is nearsighted, they may not be able to see objects clearly in the distance, or they may not be able to see anything just a few meters away.

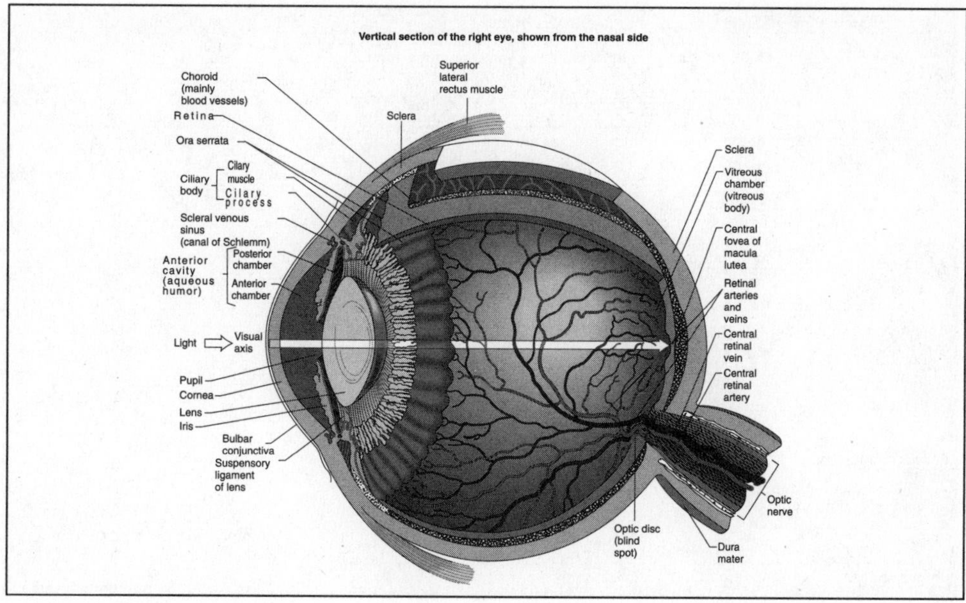

A cutaway anatomy of the human eye.

To correct for the short focal length of the lens, a diverging lens is used to separate the light rays just enough so that the light will focus on the retina.

What is **farsightedness** and what can be done to correct it?

Farsightedness occurs when the lens of the eye can see objects far away, but cannot focus in on objects at closer range. The lens of the person with farsightedness causes the image to focus on a point behind the retina. Since the retina is the only thing that can receive the information, the image is blurred at short distances. In order to correct for farsightedness, a converging lens is used to bring the light rays closer together and focus at the point where they hit the retina.

What allows nocturnal animals to **see better in the dark** than humans can?

There are four main reasons why some animals can see better than humans can at night. The first reason is that their eyes, relative to body

Eyes of a great horned owl, a nocturnal animal.

size, are larger and can gather more light than human eyes can. More light results in a clearer image.

The next reason has to do with the rods and cones in the nocturnal animal's eyes. Cones are used for detail and work best in bright light. A nocturnal animal has little need for cones and therefore has more room for the rods that detect general information such as motion and shapes.

The third reason why nocturnal eyes excel in the absence of light is due to something called the tapetum. The tapetum is a membrane on the far side of the retina that reflects light back through the retina to double the retina's exposure to light. The light that is not absorbed by the rods is sent back out the pupil and can be seen in the animal's eyes when bright lights are shown in them.

Finally, many nocturnal animals have slit pupils that allow for a quicker response when opening and closing the eye. Therefore, the eye can open extremely wide at night, while during the day it allows only a small amount of light into the eye.

CAMERAS

What causes **red-eye** in photographs?

Red-eye occurs when a flash is used because there is not enough light for a good exposure. Under normal conditions, in order for enough light to enter the eye, the pupil dilates so that an image may be formed on the retina. But when a flash is fired, the pupils are not expecting the bright light and do not have a chance to constrict. As a result, a large amount of light enters the eye and reflects off the red retina in the back of the eye. The redness on the pupil is actually the reflection off a person's retina captured on film.

What is used to **reduce red-eye**?

Red-eye reduction is a feature found on many modern cameras. It simply attempts to constrict a person's pupil so that not as much light can be reflected back from the retina. There are several methods of accomplishing this. One method is to have a smaller light that illuminates before the real flash; another method is to have a quick burst of five or six mini-flashes that cause the pupil to contract before the picture is taken.

TELESCOPES

Who invented the first **telescope**?

Galileo used the telescope to make observations of the stars and moons of Jupiter, and he was aghast at the number of stars that were not visible to the naked eye. Although Galileo exhibited and received recognition for being the inventor of the telescope in 1609, history shows that a Dutch spectacle maker by the name of Hans Lippershey invented the first telescope in 1608.

Based on the ideas of astronomers Johannes Kepler and Christoph Scheiner, the telescope was improved by using two convex lenses separated by huge focal lengths. These telescopes proved to be cumbersome and difficult to use until improvements were made to the quality of flint glass nearly 120 years later.

A reproduction of Galileo's telescope.

Reflecting telescopes, which use mirrors to focus light, were invented by Isaac Newton in 1668. Today, telescopes are relatively cheap; the average person can set one up in his or her back yard and gaze up at the heavens with much better equipment than Galileo or Newton ever dreamed possible.

What is a **refractor telescope**?

The refractor telescope was the first telescope ever created. It employs lenses to gather, refract, and focus light toward an eyepiece. Refractor telescopes are limited in the amount of light they can gather because of the weight of the lenses, as well as the distortion of color, called chromatic aberration, that is evident when light passes through different sections of the lens.

What is a **reflecting telescope**?

A reflecting telescope uses mirrors to gather light and focus the image. It usually consists of two mirrors: one large mirror at the end of the telescope to gather light and a smaller mirror used to direct the light to the eyepiece. Isaac Newton developed the first reflector telescope in 1668.

What are some of the **largest reflecting telescopes**?

The larger a reflecting telescope, the more light it can gather, making it more powerful. The following is a list of some of the largest reflecting telescopes in the world.

Diameter of the mirror (meters and inches)	Observatory	Location
10.00 m (394 in)	Keck Telescope at Mauna Kea Observatory	Hawaii
6.00 m (236 in)	Astrophysical Observatory	Russia
5.08 m (200 in)	Palomar Observatory	California
4.19 m (165 in)	del Roqye de los Muchachos Observatory	Canary Islands
4.01 m (158 in)	Cerro Tololo Inter-American Observatory	Chile
3.89 m (153 in)	Anglo-Australian Observatory	Australia
3.81 m (150 in)	Kitt Peak National Observatory	Arizona

Besides its size, what is unique about the **Keck Telescope** in Hawaii?

The engineering of the mirror at Mauna Kea's Keck Observatory is unique because the mirror has 36 hexagonal parts. Each section of the mirror is controlled electronically, which allows the positioning, cleaning, and other maintenance of the mirror to be performed with relative ease.

Why is it advantageous to have a space telescope, such as the **Hubble Space Telescope**?

The major advantage to having telescopes in space is that they are able to escape from light, air pollution, and the refractive qualities of the air that distorts the images when observed from Earth. Secondly, space telescopes can also pick up infrared, UV, X-ray, and gamma waves that otherwise would be difficult to receive through the Earth's atmosphere.

The Hubble space telescope.

What was wrong with the Hubble Space Telescope when it was first put into **orbit**?

An error (less than the thickness of a human hair) in the curvature of the main mirror caused major focusing problems for the Hubble Space Telescope. The huge mirror, 2.4 meters in diameter (94.5 inches), was not able to focus the light to the correct point in the telescope. NASA suffered great embarrassment for this multi-million dollar mistake.

What was done to correct for Hubble's **vision problems**?

Three years after the Hubble was placed in orbit around the Earth, a team of astronauts from the Space Shuttle Endeavor installed three tiny mirrors that would correct the focusing problems that the Hubble was experiencing. Since its repair, the Hubble Space Telescope has aided serious research into the age of the universe and the rate at which it is expanding, and has enabled observation of other stars and galaxies that previously were never seen by earthbound telescopes.

ELECTROSTATICS

Where did the term **electricity** originate?

Seventeenth-century physician and scientist William Gilbert coined the term "electricity" from the Greek "elektron," which meant amber. Amber was the material that the ancient Greek philosophers had noticed would mysteriously attract small particles after it had been rubbed with fur.

What is **electrostatics**?

Electrostatics is the study of those electrically charged particles that can be moved from place to place and then held at rest. Electrostatic laws deal with the attractive and repulsive forces found to exist between positive and negative electrical charges. In lay terms, electrostatics is sometimes known as static electricity.

Who first discovered **electrostatic charge**?

Two Greek philosophers, Thales of Miletus (600 B.C.) and Theophrastus (300 B.C.), made some observations about a peculiar substance called amber, which, when rubbed with fur, seemed to attract dust and other small particles. Nothing more was made of the fact until centuries later, when the physician to the Queen of England, William Gilbert, made some scientific observations and performed experiments with amber.

What is the difference between a **positively** charged particle, a **negatively** charged particle, and a **neutrally** charged particle?

A negatively charged object is an object that has an excess of negative particles called electrons. A positively charged object has an excess of positive particles, called protons. A neutrally charged object has an electron for every proton in an atom and therefore has no charge.

What combination of charges causes **attraction and repulsion forces**?

Unlike gravitational forces, which simply attract masses to each other, electrostatic forces can either attract or repel charges. Like charges (such as positive to positive or negative to negative) repel each other. Unlike charges (positive to negative) attract each other. A common phrase describing many human social relationships, "opposites attract," holds true for electrostatic forces.

What are some ways to observe **electrostatic forces**?

When a rubber rod is rubbed with fur, the fur transfers electrons to the rubber rod. The objects, originally neutral, are now charged. The rod, which is now negatively charged because it has excess electrons, can attract positive charges. Since the rod is attracted to the positive charges in other objects, it can pick up little pieces of paper and dust or even raise the hair on a person's head. Although paper, dust, and hair are attracted to the negatively charged rod, the objects themselves are still neutral, because they do not have a net charge, but instead have become polarized. Polarized, in electrostatics, means that the positive and negative charges have aligned themselves so that the positive charges in the paper or dust will move closest to the negatively charged rod, while the negative charges align themselves as far away as possible from the negative rod.

Rubbing a glass rod with silk will achieve the opposite effect. The glass rod becomes positively charged, while the silk receives the negative

Why does a rubber balloon that has been rubbed in your hair stick to a wall?

The attraction between a charged balloon and a wall is the result of electrostatic forces. When rubber is rubbed through human hair, electrons in the hair transfer easily to the rubber balloon. This is called charging by friction. The hair may stay up as a result of the excess positive charges repelling each other, but the negatively charged balloon is attracted to the many positive charges in the wall, which have polarized themselves with their positive charges close to the balloon and their negative charges away from the negative balloon. As long as the electrostatic force and frictional force between the balloon and the wall are stronger than the gravitational force pulling the balloon down, the balloon will remain on the wall.

excess electrons. The glass rod can still pick up small objects, but attracts the negative charges in those objects instead of the positive charges.

Why do you sometimes get a **shock** when touching a doorknob?

This annoyance happens usually on dry days after walking on carpeted floors. The friction between the carpet and your shoes causes your body to accept an excess of negative charges. When your hand approaches a doorknob, the negative charges in your hand are attracted to the positive charges in the doorknob (created by polarization), causing an electrical spark when the two charges meet.

What are some good **conductors** of electricity?

In order to be an effective conductor of electrical charge, a material must allow the electricity to move easily throughout it. Although most

Why is it important to beware of excess electrostatic buildup when working with computer equipment?

If you have ever installed a circuit board or card into a computer, the product probably was shipped in a "static-free" bag. This bag is designed to keep all excess static charge outside the bag. Many electronic circuits are sensitive to the electrostatic buildup, and can be damaged if such a charge accumulates on sections of the circuit. Therefore, when installing the circuit board, the instructions usually encourage you to neutralize yourself by touching a grounded piece of metal to discharge your body and tools.

materials can conduct electricity to some extent, good conductors, such as most metals (especially copper and silver), have many free electrons, which help move charges quite easily.

What is a good **insulator** of electrical charge?

Insulators are the opposite of conductors; insulators inhibit the movement of electricity. Effective insulators are usually non-metals, such as plastic, wood, stone, and glass. These materials have electrons that do not move as freely as those in conductors.

How is the **strength of an electrical force** measured?

All forces, including electrical forces, are measured in newtons. The formula for determining an electrical force is only slightly different from the Law of Universal Gravitation in that the electrostatic force uses charges instead of masses, and employs a different constant to make the formula function. A man by the name of Charles Augustin de Coulomb conducted several experiments in 1785 that determined the variables that define the strength of an electrical force; he determined that one needs to know the charge on two or more particles and the distance between the charges.

What is **Coulomb's law**?

Coulomb's law determines the exact strength of the electrical attraction or repulsion between two charged objects. The formula is $F = k (q_1 q_2/r^2)$, where k is the dielectric constant and always has a value of 9.0×10^9 Nm^2/C^2 (newton-meters squared per coulombs squared). The charges q_1 and q_2 represent the amount of charge of the objects being measured. Finally, r is the radius or distance between the two charged objects. A negative force represents an attractive force, while a positive force represents repulsion. By knowing the variables for the equation, the electrostatic force between any two charged objects can be determined.

What is a **coulomb of charge**?

A coulomb of charge is equal to the electrical charge of 6.24×10^{18} electrons or protons. A negative charge represents the charge of electrons while a positive charge represents the charge of protons.

What is an **electroscope**?

An electroscope is used to determine if an object is carrying a charge. It consists of two metal leaves (ranging from aluminum foil to gold) attached to a metal rod. If a charged object touches the top of the electroscope, the two leaves on the end of the electroscope will repel each other (because they both become charged with like charge), indicating that the object is charged. If the leaves do not separate, it shows that the object touching the electroscope is neutral.

Who **invented** the electroscope?

British physicist Michael Faraday developed the first electroscope in the mid 1700s and, with the help of the electroscope, developed the concept of the electric field.

What is an **electric field**?

An electric field is an area where there is an attractive or repulsive electric force between two or more charged particles. Just as there is a gravitational field on Earth—an area where objects of mass will be attracted to the Earth—there is an area where electrically charged objects will feel an electrical attraction or repulsion toward charged objects.

What is **Gauss' Law** used to describe?

Gauss' Law describes the connection between an electric charge and the strength and distribution of its surrounding electric field. The law is named after Carl Friedrich Gauss, primarily a mathematician, who applied his incredible mathematical skills to the research he conducted in astronomy and physics in the early nineteenth century.

Van De Graaff Generators

What is the **Van de Graaff generator**?

Named after its American creator, Robert Jemison Van de Graaff, the Van de Graaff generator has been the highlight of many electric demonstrations in both physics classrooms and museums around the world. The device, created in 1931, consists of a hollow metal sphere that stands on an insulated plastic tube. Inside the tube is a rubber belt that moves vertically from the base of the generator to the metal sphere. The rubber belt moves negative charges up the tube and into the metal sphere. Metal combs are used to capture the charges and distribute them around the exterior of the metal sphere. The huge build-up of negative charges on the sphere can reach anywhere from several hundred thousand volts to several million volts.

When people **hold on** to the Van de Graaff generator, what happens to them?

If someone were to hold on to the generator while it is charging the metal sphere, the electric charges accumulating on the sphere, which

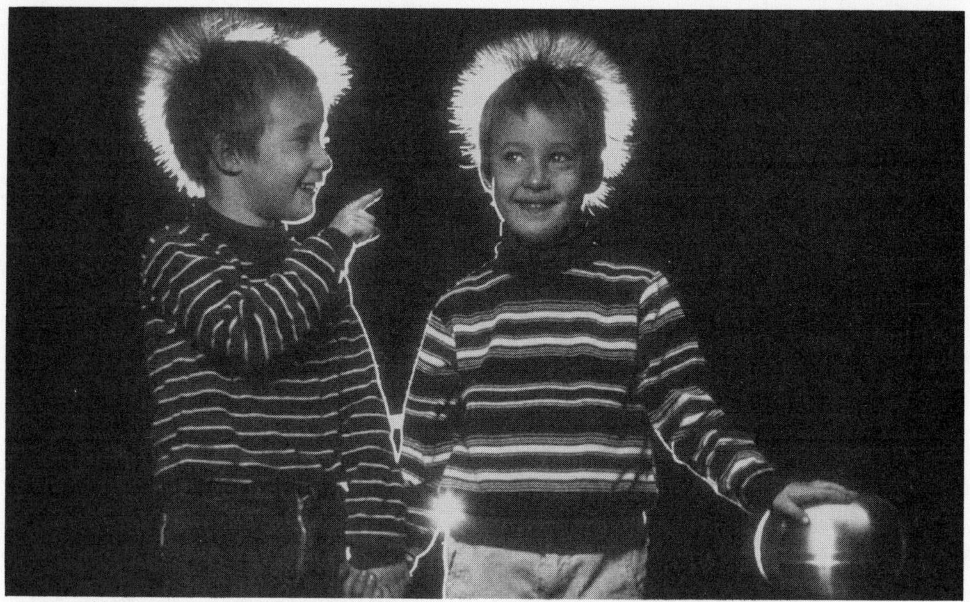

The hair-raising effects of a Van de Graaff generator.

want to repel each other, would travel onto the person's body. Eventually, when the person's body is covered with electric charge, their hair may stand up on end because the electric charges on the hair repel each other. This does not hurt the person, because there is not enough current flowing through their body to cause any harm.

What happens if the Van de Graaff is **charging** and someone approaches it?

The sphere on the Van de Graaff is a conductor surrounded by an insulator (the air). Electric charges on the sphere have a strong compulsion to leave the metal sphere, but are unable to so because of the insulating properties of the air. However, when an object is positively charged or has a different voltage potential than the sphere and comes close to the Van de Graaff generator, the negative charges jump the air gap to join the positive charge. When someone approaches a Van de Graaff generator while it's charging, the person will receive a shock and see a mini-lightning bolt between an extremity, such as a finger or nose, and the Van de Graaff generator. Although the shock can be painful, it causes no harm to the person. These scenarios are using typical Van de Graaff generators found in physics classrooms, not the huge generators found in research laboratories or museums.

There are hundreds of thousands of volts on the metal sphere of the Van de Graaff generator; how much charge is **inside** the sphere?

Zero. When negative charges leave the rubber belt, they move immediately to the outer perimeter of the sphere. Negative charges like to be as far away from each other as possible; this is why they move to the outermost surface of the Van de Graaff generator.

What is a **Faraday Cage**?

A Faraday Cage, named after British physicist Michael Faraday, is a cage or metal grating that can shield electrical charge. Charges gather on the outer shell of the cage, for they are repelled by one another and can be further from each other if they are on the outside of the cage. This results in a neutral charge within the Faraday Cage. The metal sphere of a Van de Graaff generator is a Faraday Cage. Cars and airplanes can be Faraday Cages as well, and may provide some protection from lightning during an electrical storm.

LEYDEN JARS

What is a **Leyden jar**?

A Leyden jar, the basis for the modern day capacitor, is a glass container with a rubber stopper that can store electrical charge. It consists of two conducting plates, usually aluminum foil, separated by a dielectric or insulator, usually glass or plastic. The inside sheet of aluminum foil is placed inside the jar and charged. The outer coating of aluminum foil, on the outside of the glass jar, is connected to the ground so it accumulates the opposite charge. When finished charging, the Leyden jar can be carried around, fully charged, until it is needed.

To discharge the Leyden jar, a metal wire is connected from the outer, positive coating to the inner, negative coating of the Leyden jar. When the connecting wire approaches the inner coating, a relatively large spark is generated, and the Leyden jar becomes neutral once again.

A Leyden jar.

Who **invented** the Leyden jar?

In the 1740s, Dutch scientific instrument makers Ewald von Kleist and Pieter van Musschenbroek created the first Leyden jar, but they had no idea of its potential. A jar was used because electricity back then was thought to be a fluid, and jars were usually used to store fluids.

The original Leyden jar had a metal nail and water on the inside of the glass jar. Musschenbroek held the outside of the jar, and the nail received one charge from an electrostatic machine while Musschenbroek accumulated the opposite charge from the ground. When Musschenbroek touched the nail with his free hand, he received a massive shock! This was the first artificially created shock, and Musschenbroek stated that he would never perform such an experiment again. However, the next day he changed his mind and was back working on the Leyden jar.

What accomplishments did **Benjamin Franklin** make with the Leyden jar?

Benjamin Franklin made many contributions to electricity, but one of the more interesting and daring experiments that he performed involved the Leyden jar. Franklin introduced the idea of positive and negative charge and realized that electric fields existed between each side of the glass insulator. Finding an electric field between the two plates of the Leyden jar was a big step in the field of electrostatics and electric fields. Mr. Franklin attempted to capture the electricity from lightning in his

Leyden jar. He was somewhat successful, for he was able to capture some of the lightning's electrical charge in the jar and created a mighty spark when connecting the outside plate with the inside plate. The fact that Franklin only captured some of the lightning in the Leyden jar was a blessing for him, for the next person who attempted to capture the electricity from lightning died during his attempt. Franklin learned early on that electricity was not something to fool with.

What would the **Leyden jar** be used for?

In the late eighteenth and nineteenth centuries, people attempted to use the Leyden jar in a variety of ways. Some felt that it could cure medical ailments, and many doctors used the jar as primitive electroshock therapy. Others used it as a demonstration device and for entertainment purposes. Still more people felt that it could be used in cooking. Try cooking a turkey with an electrical spark!

What is the **modern-day version** of a Leyden jar?

If someone needed to store electrical charge today, they would not be seen walking the streets with a Leyden jar. Instead electrical charge is stored in a capacitor. Each capacitor, depending upon its design, has the capacity to store a particular amount of electric charge. Capacitors, like Leyden jars, consist of two conducting plates and an insulator placed between them to create an electrical field. To discharge, the two plates are connected with a wire, creating a sudden electrical discharge.

CAPACITORS

What are some **capacitors** used for?

Capacitors are used to store large amounts of electricity for later use, and are found in all kinds of electrical circuits. Camera flashes rely primarily on a capacitor. Typically, the batteries in a camera would not be able to produce enough of an electrical current to light a flash. To solve

What is the whining sound that is sometimes heard after a camera flash is fired?

After a capacitor has discharged, the camera needs to prepare itself for another flash. Therefore, the capacitor charges up again and is sometimes accompanied by a whining sound from the electrical circuitry inside the camera.

this problem, a capacitor is used to store plenty of electrical charge, so when a surge of electrical current is needed, the capacitor discharges, sending lots of current to the flash.

What are the **units of electrical capacitance**?

Capacitance is measured in Farads, a unit named after Michael Faraday. Capacitance is the ratio of the charge stored to the voltage difference on the plates of the capacitor, or $C = Q/V$, where C is the capacitance, Q is the charge stored, and V is the voltage difference on the plates.

LIGHTNING

What is **lightning** and how is it **created**?

Lightning is an electrical discharge that occurs between charged areas of thunderclouds or between thunderclouds and the ground. The separation of charge in a cumulus nimbus cloud forms a huge electrical potential difference between the top, positively charged region of the cloud, the bottom, negatively charged region of the cloud, and the induced, positively charged ground. When enough electrical potential builds between the positively induced ground and the negatively charged underside of the cloud, the charges collide and a massive flash of light is seen.

How do **thunderclouds and the ground** become **charged**?

Most physicists believe that charges separate when ice particles and water droplets rub against each other within the cloud. The separation of charge gives the upper section a positive charge, while the lower section of the cloud receives a negative charge.

The ground is charged by a method called induction. The build-up of negative charges on the underside of a thundercloud attracts the positive charges in the ground. The ground therefore sends most of its negative charge further into the ground, leaving a positively charged surface.

How do the charged regions of clouds and the ground act as a **giant capacitor**?

A capacitor consists of two conducting plates with opposite charge separated by an insulator. When a wire is connected between the two plates, a large electric current flows, creating a spark.

The charged regions of the clouds act as conducting plates while the air between them acts as the insulator. The same thing occurs between the lower section of the cloud and the ground. The air between these sections acts as the insulator, but when charge is able to sneak its way through the air and come in contact with the opposite charge, a tremendous current flows, creating a bright flash of light.

Does lightning always **strike the ground**?

Although lightning between the Earth and clouds is what most people think of as lightning, the most common type of lightning occurs inside and between thunderclouds. It is usually easier for the charges to jump between the clouds than it is for the charges to jump from the clouds to the Earth. As a result, only one quarter of all lightning strikes actually strike ground.

Lightning on the metro Tucson area skyline.

How does the charge pass through the **air**?

When there is enough electrical potential difference between the cloud and the ground, a negatively charged "step leader" will fall from the cloud and make its zigzagged and often branched trip to the ground. The positive charges on the ground, sensing the attractive force from the step leader, will emit streamers of positive charge from tall objects, such as trees, buildings, and towers. When the step leader and the streamer meet, the discharging of the cloud and ground takes place. The lightning first travels from the point of contact to the ground, and then each successive branch of the step leader discharges to the ground. Therefore, in a sense, the lightning takes place in several upward steps, but the current flows down to the ground.

What is the average **voltage, current, and duration** of a lightning flash?

The potential difference or voltage just before a flash of lightning can be tens of millions of volts. The current in a typical bolt of lightning is any-

How often does lightning strike?

On average, it is estimated that there are about 100 occurrences of lightning throughout the world in any given second. Since most lightning occurs within clouds rather than from a cloud to the ground, this means there are approximately 25 strikes to ground every second.

where from 25,000 to 30,000 amperes of electricity. The entire flash of lightning, however, only lasts about a quarter of a second.

How many people are **killed or injured** by lightning?

Of the 40 million lightning strikes per year in the U.S., 400 of those strikes hit people. Half die as a result of their strike, while the rest often sustain serious injuries.

Where in the United States does lightning occur most **frequently**?

A section of Florida known as "lightning alley" is a 60-mile wide hot-spot of lightning activity in the United States. This region experiences an average of 90 days with lightning strikes each year.

Safety Precautions

Is it true that lightning **never strikes** the same place **twice**?

This is absolutely false. The Empire State Building in New York City is just one example of where lightning has struck more than once. In some thunderstorms, the tower on the Empire State Building has been hit several dozen times.

Why is a **car** often the best place to be when lightning strikes?

It is not the rubber tires! Many people think the rubber tires of a car provide insulation from the lightning striking the ground. If this was the case, wouldn't riding a bicycle do the same thing? The real reason why a car is a safe place to be when struck by lightning is because most cars have metal bodies, which act as Faraday Cages, keeping all the electrical charge on the outside of the car. Since the charge is kept on the outside of the vehicle, the person sitting inside the car is kept perfectly neutral and safe. It is the shielding of the metal car body, and not the rubber tires, that protects people in automobiles.

What happens to an **airplane** when it is struck by lightning?

Airline pilots tend to avoid thunderstorms, but when a plane is struck by lightning, the passengers inside the plane are kept perfectly safe, for they are inside a Faraday Cage, which shields them from the massive electrical charge.

Studies were performed by NASA in the 1980s in which they flew fighter planes into thunderstorms to see how the planes would react to lightning. The scientists quickly found that the planes actually encouraged lighting, because the planes caused compressions in the electric field of the cloud, which in turn caused the lightning to hit the plane's metal body.

What are some things you should **do** if caught in a **lighting storm**?

The safest place to be during an electrical storm is inside a building (where you should stay away from electrical appliances such as the phone and television, as well as all plumbing and radiators) or car, but if you are unable to shield yourself in this way, the following precautions should be taken:

Crouch down on the lowest section of the ground, but do not let your hands touch the ground. If lightning strikes the ground, the electricity spreads out sideways and can still reach you. If only your feet are on the ground (especially if you're wearing rubber-soled shoes), this might limit the amount of electricity that passes through your body. If you must lay down because of an injury, try to roll up into a tight ball.

Take off and move away from all metal objects unless they act as Faraday Cages (refer to the question about Faraday Cages).

Move away from isolated and tall trees.

Avoid the tops of hills or mountains and open areas such as water and fields.

If out on a lake or on the ocean, get back to shore as quickly as possible. If that is not practical, get down low in the boat and move away from any tall metal masts or antennas.

Lightning Rods

Why are **lightning rods** effective in keeping **tall trees and homes** safe from lightning?

Lightning rods are pointed metal rods that rise above a tree or rooftop to protect the object. The rod, connected to the ground by a metal wire, both encourages and discourages a lightning strike. The rod discourages the lightning strike by "leaking" positive charges out of its pointed top to satisfy the need for positive charge in the clouds. If the rod cannot leak out enough charge to satisfy demand, the step leader from the cloud is instead attracted to the rod, and a flash of lightning occurs. Therefore, the rod attempts to discourage lightning, but if it cannot satisfy the negative charge, it attracts the lightning to the rod instead of the tree or house.

If the lightning rod doesn't have a good connection to the ground through the grounding wire, however, it can increase the danger to the building. Often these heavy grounding wires come loose from the lightning rods, and if the rod is then hit by lightning, the current will flow along the surface of the building to ground and could cause a fire. The rods can become disconnected from lack of routine maintenance; it is wise to check these connections on a regular basis.

Why shouldn't you stand under a large tree during a thunderstorm?

During thunderstorms, many people stand under trees in an effort to stay dry. However, this can have dire consequences. In the spring of 1991, a lacrosse game at a Washington, D.C., high school was postponed after lightning was observed in the sky. Over a dozen spectators ran for shelter under a tall tree to protect themselves from the rain. A few seconds later, lightning struck the tree, injuring twenty-two people and killing a fifteen-year-old student.

Trees are tall points where positive streamers can be released into the air and attract the step leader, causing lightning. By standing under a tree, holding an umbrella, swinging a golf club, or batting with an aluminum bat, people are making themselves part of a lightning rod.

Who **invented** the lightning rod?

It was Benjamin Franklin who invented the lightning rod to protect area houses and tall trees from being destroyed by devastating lighting bolts.

ELECTRICAL CURRENT

Who was **Luigi Galvani**?

Italian physiologist Luigi Galvani (1737–1798) is known for his accidental discovery of electrical current. Galvani was not looking for a way to create electrical flow; in fact, sustained electrical flow was not even considered a possibility in the 1780s and '90s.

Galvani proved that when two charged metal probes came in contact with a dead frog's leg, the leg would twitch. Galvani felt that the result of this experiment was more of a physiological discovery than a physical or electrical discovery.

How was Galvani's experiment helpful to **Alessandro Volta**?

Alessandro Volta, a colleague of Galvani's at the University of Bologna, proved that the twitching was not only a biological phenomenon, but that the twitching was a result of some of the electric flow caused by the two charged metals touching the frog's muscle. The muscle, according to Volta, acted as a conductor for the flow of electricity. Volta was successful in proving that it was not the nerve that produced the current, but instead that it was the two different metals, using the muscle as a conductor, that produced the current.

What was the **most important result** of Galvani's experiment?

The knowledge that Volta obtained from Galvani's accidental experiment led to the development of the voltaic pile, which provided a method of creating a sustainable electric current.

What is the **voltaic pile**?

A voltaic pile was the first electric battery, which Alessandro Volta created in the late eighteenth century. In between the metal electrodes of silver and zinc, Volta placed cardboard dipped in salt water to separate the metal and conduct electricity. He found that electrons moved from the zinc electrode through copper wire and to the silver electrode. The development of this voltaic pile became the predecessor to the dry cell battery of today. The discoveries of Galvani and Volta in the late 1700s helped move scientific research from static electricity into the new field of electrical current.

What is **current**?

Current is defined as the flow of positive charge. It is measured in amperes (or "amps") and indicates the amount of charge that passes through a wire in one second. In order to produce current, a source of electricity (such as a battery or generator) is needed.

What is **voltage**?

Named after Alessandro Volta, "voltage" is the potential difference in an object or battery. To produce current, a difference in electrical potential is needed to get the electricity to flow. This concept is analogous to the flow of water; electricity flows from an area of high potential energy to lower potential, just as a river flows from high elevations to lower elevations. In an electrical circuit, the battery often is the source of potential difference. The positive terminal of the battery contains the high potential, and the current will flow through the circuit to the negative or low-potential section of the battery.

RESISTANCE

What is **resistance to current flow**?

All objects encounter friction when moving. Electrons are no different, but we refer to the friction that electrons encounter as resistance. Resistance slows down the electric charges and in the process causes wires or any other conductor of electricity to heat up. When resistance is high, the temperature can increase enough to operate light bulbs or toasters. On the other hand, resistance can be quite detrimental as well, and burn up certain electrical devices when there is too much resistance.

What **four factors** determine the **amount of resistance** that a wire will have?

The resistance of a wire depends upon the following:

The length of the resistor (the longer the wire, the more resistance).

The cross-sectional area of the resistor (the thinner, the more resistance).

The temperature of the resistor (the warmer, the more resistance).

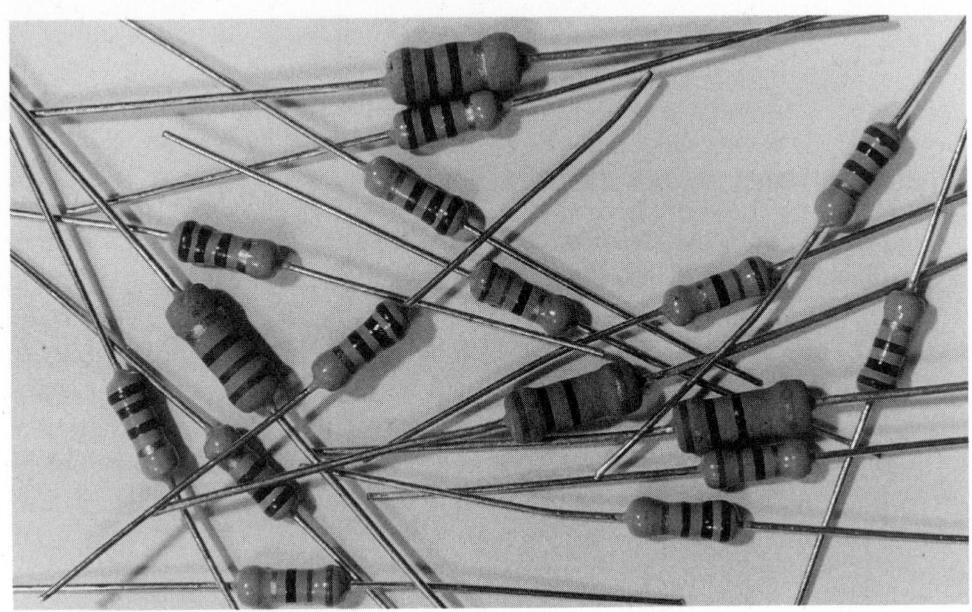

Color-coded resistors.

The properties of the material (the fewer the number of free electrons, the more resistance).

What do **color bands** on resistors represent?

If you have ever opened up an electrical device, you have probably seen little cylindrical devices that have four color bands around them. These color bands represent the particular resistance value of the resistor. The bands are extremely important to electrical engineers when they need to speed up or slow down current flow in their circuits.

Each color band on a resistor represents a particular number or multiplying factor for that particular resistor. The first two bands represent the first and second significant digit. The third band represents the multiplying factor, while the fourth band defines the accuracy or percent error for the resistor. For example, if a resistor had orange (which represents 3), blue (which represents 6), yellow (which represents 10^4), and gold bands (which represents 5%), it would have a resistance of 36×10^4, with a 5% margin of error.

SUPERCONDUCTORS

What is a **superconductor**?

A ceramic superconductor.

Superconductors allow electrical current to travel without resistance. All electrical systems would run much better and more cheaply if electricity could travel without resistance. Superconductors are elements that have no electrical resistance when cooled past their low critical temperatures. Some elements used for superconductivity include aluminum, lead, and niobium. However, developments have been made in ceramics that could make many materials superconductors without reducing their temperatures to the low-critical point. This is significant, for not as much energy would be needed to create a superconductor.

Who **discovered** superconductivity?

The creation of materials without resistance was thought to be impossible, but a Dutch physicist by the name of Heike Kamerlingh Onnes proved it was possible in 1911. Onnes lowered the temperature of different metals, including mercury, close to absolute zero. He then measured the electrical resistance of the materials at such low temperatures and found that mercury, at only 4.2 Kelvin (-277.2°C), had zero resistance to electrical current.

What **technologies** have developed as a result of superconductivity?

Technologies such as magnetic resonance imaging (MRI)—a method of viewing the human body without using harmful radiation, geological sensors for locating underground oil, and particle accelerators, which attempt to reveal the fundamental structure of matter by smashing sub-

Who won the Nobel prize for their work in superconductivity?

Three American physicists, John Bardeen, Leon N. Cooper, and John R. Schrieffer, explained why superconductivity occurs in particular materials. Their development of the BCS theory for superconductivity helped them win the Nobel prize in 1972.

Fifteen years later, two other physicists won the Nobel prize for discovering superconductive materials that achieved zero resistance at temperatures thought to be too high for superconductors. Physicists Georg Bednorz and Alex Müller of IBM found that a ceramic substance called lanthanum barium copper oxide became a superconductor at 35 Kelvin. This was a much higher temperature than any physicist thought possible at the time.

atomic particles together, have been created as a result of superconductor technologies.

What would **future advancements** in superconductivity mean for science?

To further improve upon superconductivity technology would have significant effects on technologies that rely on electricity and magnetism. Within the not-so-distant future, superconductors may be used to generate, transmit, and store electricity. Superconductors may also be used to detect electromagnetic signals, protect communities from power surges, and help develop higher quality and faster cellular telephone technology. Technologies a bit further into the future include the development of superconducting magnetic levitating trains and super-fast computers.

Does voltage shock you?

Signs around powerplants and breaker boxes often state, "CAUTION: High Voltage Area." It is not voltage that can hurt you; it is the electrical current that flows through your body that can produce serious and sometimes fatal consequences. The Van de Graaff generator generates hundreds of thousands of volts, but produces such a low amount of current that the sparks it emits only cause muscles to tingle.

OHM'S LAW

What are the **units and symbols** used for current, voltage, and resistance?

Term	Unit	Symbol for unit
Current (I)	Amperes (or amps)	A
Voltage (V)	Volts	V
Resistance (R)	Ohms	Ω

How do these three values **work together**?

Voltage, current, and resistance form the basic law of circuits. In the early 1800s, Georg Simon Ohm, a German physicist, developed the law that would one day bear his name. He discovered the relationship between voltage, current, and resistance, and created the following formula based on their relationship: Voltage = Current × Resistance.

How much **resistance** do our **bodies** have to electrical current?

On average, the human body has an electrical resistance between 300,000 and 700,000 ohms, depending upon the particular characteris-

An electric chair.

tics of our bodies. It is relatively difficult to give ourselves a fatal shock, unless we reduce our resistance to electrical flow. For instance, if we make ourselves wet, we reduce our resistance all the way down to a couple of hundred ohms. In this state of low resistance, it would be quite easy to electrocute ourselves.

How much **current** is needed to cause **pain and death**?

Very low amounts of current passing through our bodies can cause pain and even death. However, the resistance of the human body makes it extremely difficult for electricity to pass through with high current. Also, the end result of current flowing through one's body also depends on how the path the electricity takes through the body. For example, if the current passes through the heart or brain, severe damage or even death can occur; if the current passes through the legs or arms or along the exterior of the body, there would be less damage. In general, however, to create a minor pain sensation, a current of only 0.005—0.010 amps is needed. A current of 0.07 amps could kill a person. Again, the high resistance of the body usually prevents great damage from occurring.

How much voltage does the **electric chair** place across a person?

Supposedly, the electric chair allows a person to fall unconscious and die a painless death. Such a statement might seem a bit strange, given that an electrical shock is typically quite painful and can be extremely dangerous. The general practice for the electric chair is to place a 2000-volt source across the person in the chair. This is not to say that a lower voltage, however, cannot cause pain or death.

Given the average human body's resistance of 500,000 ohms, a 2000-volt potential difference would produce a meager .004 amp current, enough for pain only. What is done to **increase the current**?

Since it is current flowing through the body and not the voltage that electrocutes someone, several things are done to ensure that the prisoner will not survive the experience in an electric chair. In order to reduce the resistance, two electrodes dipped in salt water are placed on the prisoner. The salt water helps conduct the electricity by reducing the resistance of the skin to approximately 5,000 ohms, which allows the now 0.4-amp current passing through the person to kill him or her in a matter of minutes.

Do **electric eels** really emit electrical charges to capture their prey?

Electric eels do indeed set off electrical pulses to stun and even kill their prey. These eels have special nerve endings bundled together in their tails that can produce 30 volts in small electric eels, to 600 volts in larger eels. Besides using the electrical shocks for hunting, the eels produce

An electric eel.

a constant electrical field for use in navigation and self-defense. Most people do not have to worry about encountering electric eels, however. This variety of eel is native only to the rivers of South America.

Why don't **birds or squirrels** on power lines get electrocuted?

In order to get electrocuted on such well-insulated wire, a bird would have to be in contact with objects that had two different voltages. For example, if the bird were touching the wire (high voltage) and the ground (low voltage), a large amount of current would pass through the bird, electrocuting it. Humans could also hang from a power line, as long as they were not close to or touching something with a different voltage such as a another wire, telephone pole, ladder, or the ground. It is not recommended, however, that you test this; just take our word for it.

Why do many electricians work with **"one hand behind their back"**?

When working on difficult high-voltage circuitry, many electricians like to place one hand behind their back because this way there is little

Birds on electric wires do not get electrocuted because the birds are touching only the wires and not any other object of different voltage.

chance for each hand to touch objects of different electrical potentials and cause a major spark and shock. In addition, if an electrician has one hand behind his or her back, it is difficult for a current to take a path between the hands and directly through the heart—even a small current through the heart can kill quickly.

WATTS AND KILOWATTS

What's a **watt**? What's a **kilowatt**?

A watt is the unit for power. Specifically in electricity, it is the unit for electrical power and is often found on light bulbs and other devices used in electrical circuits. To determine the power usage of a component in an electrical device, physicists apply the formula $P = I \times V$ (power = current $\times$ voltage). A kilowatt is a thousand watts of power.

Why is a 100-watt bulb **brighter** than a 25-watt bulb?

The wattage indicates the amount of power that the bulb consumes. When plugged into a normal 110 or 120 volt outlet, more current passes through the 100-watt bulb than the 25-watt bulb, and the more current that passes through the bulb, the brighter and more powerful the bulb.

Kilowatt vs. **kilowatt-hour**—what's the difference?

A kilowatt is the unit used to describe the power that a particular device or a building uses. It measures the rate at which the energy is being consumed by multiplying the current the appliance or bulb needs by the voltage across it. The electrical companies do not care how quickly you use energy, just how much of it you use. Therefore, the utility company charges you for the number of kilowatt hours of electricity you use; that is, the total amount of energy used, not the rate at which the energy is used.

For example, a 100-watt light bulb uses 100 watts (or 0.1 kilowatt) of power. If that light bulb were left on for an entire month, the energy that the bulb consumed would be 0.1 kilowatts × 24 hours × 30 days, which equals 72 kilowatt-hours of energy. If a kilowatt-hour costs $.12 per kilowatt-hour, the bill for that one light bulb would be $8.64 per month.

CIRCUITS

What is needed to **create a circuit**?

Ohm's law refers to three different variables that are needed in a circuit. The first is a voltage supply, or potential difference in the circuit. This can usually be achieved by connecting a battery, wall outlet, or some other power supply to the circuit. The next variable is current. In order to have current flow, wires are needed to connect the power supply to the resistors and back to the power supply. Finally, resistance is needed. Resistance to the flow of electrons can be achieved with wires, electrical devices, and even the power supply itself.

What is a short circuit?

A short in a circuit occurs whenever there is not enough resistance in the circuit. Shorts can occur when two wires touch and the resistor in the circuit is bypassed. This increases the current and leads to overheating in the circuit and electrical fires.

What is a **closed circuit** compared to an **open circuit**?

By using a switch in a circuit, the operator can open (turn off) the circuit or close (turn on) the circuit. In order to have electricity flow, a circuit cannot have any gaps. It must form a complete loop from the positive side of a power supply to the negative side of the power supply—otherwise it is an open, and therefore useless, circuit.

What can be done to **prevent** a short circuit?

When a short occurs, a surge of current causes the temperature to rise. When the current gets too high, the circuit can be turned off or "opened" by flipping a "breaker" or by melting a thin wire in a fuse.

AC/DC

What important contribution to electricity did **Nikola Tesla** make in the late nineteenth century?

Nikola Tesla, a former employee of Thomas Edison, was a key figure in the development of the first system to produce and send alternating current for electric power. Some of his other accomplishments include the development of generators and the Tesla coil, used as a transformer for electromagnetic radio communication.

What is a **DC** circuit?

A DC or direct current circuit allows electrons to travel only in one direction throughout a circuit. Most DC circuits consist of a battery or direct current power supply, wire, and varying types of resistors.

What is an **AC** circuit?

Instead of moving electrons in one direction through a circuit, AC, or alternating current, vibrates the electrons in the circuit back and forth, sixty times a second. An alternating current is usually found in wall outlets in buildings. Most of our electrical appliances run on alternating current.

Why is **AC preferred** over DC?

Back when the standards where made for electrical transmissions, the debate between AC and DC was hotly contested. The outcome was AC because of the problems that occurred when trying to send electricity over long distances. It was easy to make high AC voltages with transformers, but there was no easy way to make a similar transformer for DC voltages for several decades. Therefore, the AC won because of the development of the transformer.

SERIES/PARALLEL CIRCUITS

What is a **series circuit**?

A series circuit consists of electrical devices such as resistors, capacitors, batteries, and switches arranged in a single line. There is only one path for the electricity to flow through, and if there is a break anywhere in the series, none of the devices will work.

Why are Christmas lights parallel circuits?

For years Christmas lights used to be wired in series to save on the cost of wire. The down side was that, inevitably, every year one bulb would go out on the string of lights, which opened the circuit and caused all the bulbs to go out. Recently, more and more manufacturers have made lights that are wired in parallel, so if one light burns out, the others remain lit.

What is a **parallel circuit**?

A parallel circuit allows the electricity to flow through different branches. For example, three bulbs can be placed in separate branches of the circuit. If one wire was severed, the bulb in line with the break would not light, but the other bulbs would still have current flowing through them and would still work.

What happens to a **series** circuit if **more bulbs** are added?

If more light bulbs or other resistors are placed in a series circuit, the electricity encounters more resistance and all the light bulbs dim.

What happens to a **parallel** circuit if **more bulbs** are added?

A parallel circuit has the advantage of breaking up current into separate branches. If another branch is added with another bulb, this allows the current to take another path, and cuts down the resistance. This is similar to adding another lane on a highway; the more lanes that traffic can have, the less congestion. Therefore, when a bulb is added to a parallel circuit, the bulbs do not dim.

Are series or parallel circuits used in our **homes**?

Overall, the circuitry in homes and offices are wired in parallel. If the wiring were done in series, every time someone flipped a switch, the entire house would either turn on or off. To avoid the inconvenience of wiring each device to its own circuit, a home is wired in parallel. Therefore, if someone turns on an additional light, the rest of the lights do not dim.

The only downside to having a house wired in parallel is that if too many appliances are running, the resistance drops and the current flows too much, flipping a circuit breaker or burning a fuse to close the circuit to avoid fires or shocks. If this happens often, an electrician needs to add another circuit in that section of the house.

ELECTRICAL OUTLETS

Many outlets seem to have **three holes**—what is the purpose of each hole?

The top two holes may look identical, but they have two different functions. The slot connected to a black wire, usually on the right side of the outlet, is the hot wire. The hot or live carries the 120 volts needed to run an appliance, and is connected on one side of the appliance's circuit. The left slot, attached to white, is the neutral wire. The neutral wire is connected to the other side of the appliance's circuit, and has a voltage of 0. Remember, in order for electricity to flow, there must be a potential difference across the device. The hot and neutral wires provide this difference. The third, round hole on the bottom of the outlet is attached to green for ground.

What if the tool or appliance has a **three-prong plug** but you only have access to a two-slot outlet?

Do not use the appliance if you do not have the proper outlet for the device. Some people actually cut off the grounding wire or plug the device into a three-to-two-hole adapter. This defeats the safety feature of

What is the purpose of a grounding wire?

The ground wire is a safety device and should never be bypassed. In the event of a short circuit occurring in an appliance or tool that a person is using, the dangerous current, instead of passing through your body when you touch the appliance, gets sent safely to the ground through the grounding wire. The current "chooses" the ground wire over your body because the ground wire has a much lower resistance than you do.

the third prong. That grounding wire was put there for a reason, and to cut it off or eliminate it from the circuit could be dangerous.

What is the **little green wire** or plate on the three-to-two adapter?

The green wire attached to some adapters is the grounding wire. Since the adapter is circumventing the ground prong, an alternate means of grounding is needed. If the screw on the outlet plate is grounded, the green wire on the adapter should be attached to it. This way, if there is an electrical short, the current can still flow through the grounding wire.

What is a GFI, or **Ground Fault Interrupt** outlet?

A ground fault interrupt outlet is now required by building codes for outlets in the general vicinity of sinks or any other environment where water could encourage an electrical shock. The GFI can detect a loss of current, meaning the current has found another path to take. When this occurs, the GFI shuts off the circuit within milliseconds of a short circuit.

A ground fault interrupt (GFI) outlet (see previous page).

Why is it **dangerous** to operate electrical devices in **bathtubs, showers, and over full sinks**?

Although water reduces the resistance of the human body and makes it more susceptible to electrical shock, it is the plumbing that is the main hazard. Take, for example, a person who likes to listen to a plugged-in radio while sitting in the bathtub. If the radio happened to take a dive into the tub, the water would short circuit the radio and send electricity throughout the water. But more importantly, metal plumbing from the tub is connected to ground, and the grounding path would cause a wonderful flow of electricity that would be extremely difficult to stop. Unfortunately, this translates into a bad day for the bather; batteries are well worth the cost in this case.

Does a bathtub have to be full of **water** to experience an **electrical shock**?

No—the water is not crucial to experience an electrical shock. As long as a person is in contact with part of the device that carries electricity and a grounding source like metal plumbing, an electrical shock can easily occur. Ground fault interrupt receptacles in bathrooms today are designed to prevent such events from occurring, but it is best to minimize any contact between electrical appliances and plumbing.

The United States uses a **120-volt system**, while European countries use a **220-volt system.** Why?

The United States was the first country to establish widespread use of electricity for the public. At the time it was implemented, a 120-volt system seemed to provide enough voltage for users, yet was not enough to burn up electrical devices such as light bulbs. When electrical wiring was installed in European countries some years later, technological advancements allowed those devices to safely run off of more voltage. Therefore, the standard became 220 volts for Europe, while the United States remained at 120 volts.

THE LIGHT BULB

Who **invented** the electric light bulb?

The first person to generate light from a wire filament was a British chemist by the name of Sir Humphry Davy. Davy was known for his work on electric arcs in the early 1800s, and his breakthrough discovery was the first electric light. His light was created by passing electricity through a very thin piece of platinum wire. The light that Davy pro-

Thomas Edison and his electric light bulb.

duced was by no means practical or even useful, but it did pave the way for others to create electric lamps.

What advances did **Thomas Edison** make in the field of electrical lighting?

Although he did not create the *first* electric light, Edison did develop the first *practical* electric light, in the sense that it would last for hours more than any other bulb at the time. In 1878, Edison used a carbon-filament bulb placed inside a vacuum sealed glass container to create his electric light.

Did Edison make any other **advancements in electricity**?

Thomas Edison made many contributions to electricity and the use of the electric light bulb, including his plan to set up a parallel circuit for electric light bulbs. If bulbs were wired in series, the bulbs would dim every time a new bulb was added. In parallel, consumers could add more bulbs to the circuit without decreasing the intensity of the other bulbs.

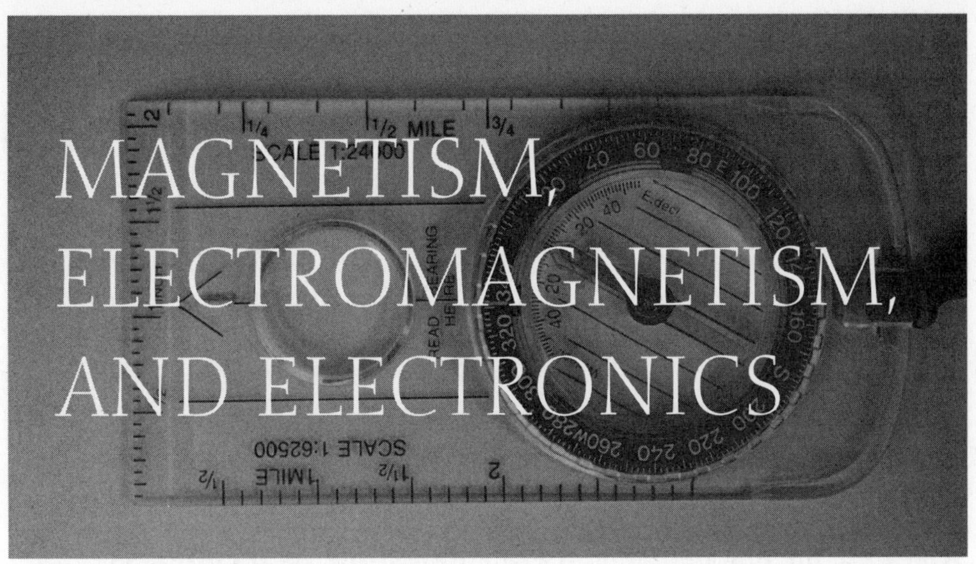

MAGNETISM, ELECTROMAGNETISM, AND ELECTRONICS

What is **magnetism**?

Just as there is an electrical force between charged particles, and a gravitational force between objects of mass, there is a magnetic force between magnetic poles. The poles of a magnet are the north pole and the south pole. Children learn early when playing with magnetic toys that north poles and south poles are attracted to each other, while similar poles repel.

Magnets consist of tiny, aligned magnetic domains inside materials such as iron. Magnetic domains are atoms that have electrons, each with a north and south pole, spinning in the same direction. It is the overall alignment of the domains that give an object its magnetic properties. In other words, they are magnets themselves that make up a larger magnet.

In addition, magnetic forces can also be generated by using electricity. Around any current-carrying wire is a magnetic field. The magnetic field is created by the uniform motion of electrons in the wire. Therefore, it can be said that they key to magnetism is the uniform motion of electrons.

What happens when a magnet is **cut** in **two pieces**?

If a magnet were severed, the domains within the magnet would remain aligned, thereby creating two magnets, each with its own north and south pole. Physicists have been searching for something called a magnetic monopole, a magnet with one pole, but quite a bit of debate surrounds the issue, and such a magnet has yet to be found.

When were magnets discovered?

People have been aware of magnetic forces ever since the ancient times of the Chinese, Romans, and Greeks. The lodestone, which contains iron ore or magnetite, was observed in these ancient civilizations to attract other lodestones.

Who were the first people to use the **lodestone as a compass**?

Although a Frenchman by the name of Petrus Peregrinus, or "Peter the Pilgrim," published some scientific writings on the usefulness and properties of the lodestone in 1269, the Chinese had been using it for navigation many years prior to Peregrinus.

How is the **Earth's magnetic field** oriented?

As seen when looking at the needle of a compass, the magnetic field of the Earth is aligned in a north-south direction. The Earth's magnetic field emanates out of the ground near the south pole, travels nearly parallel to the ground, and then turns into the Earth near the north pole.

Does the north pole of a compass point north?

When a compass is free to rotate, it experiences a torque and turns because it is attracted to opposite poles. When we look at a compass, we say the north end of the needle points north. However, the north pole of a compass needle can never be attracted to the north pole, because only opposites can attract. Therefore, what we call our north pole, up in the Arctic, actually behaves like the south pole of a magnet. In other words, the Arctic "north pole" is actually the south pole—and the Antarctic "south pole" is actually the north pole—of this giant magnet we call Earth.

What did Columbus discover as he was crossing the Atlantic in 1492?

Columbus realized that when he used a compass, the north end of the needle seemed a bit different from what the position of the stars dictated to be north. Columbus found the needle to change orientation more and more as he crossed the ocean. Columbus had discovered magnetic declination.

Why is the Earth magnetized?

No one is absolutely sure what gives the Earth its magnetic field. Some scientists feel the inner molten core of the Earth is partially metallic and generates a magnetic field through the movement of its electrical charges.

Has the Earth's magnetic field always remained **stationary**?

Through observations of iron ore in the Earth's crust and in sediments below the oceans, geologists have speculated that the Earth's magnetic field has probably reversed approximately nine times over the past 3.5 million years, and that the position and intensity will probably change again over the next several thousand years. In fact, the magnetic north pole has moved as much as 800 kilometers toward the geographic north pole since its discovery in 1831.

What is **magnetic declination**?

Magnetic declination is the angular difference between the magnetic north pole and the geographic north pole. The geographic north pole is the location of the Earth's axis of rotation; the magnetic north pole, where compasses point, is not fixed and is therefore not located exactly on the geographic north pole. This detail was not noticed because of the relatively short distances between any two points in Europe, but became readily apparent on such a long journey as Columbus made across the Atlantic. Interestingly enough, Columbus mentioned nothing of this to his crew, for he was afraid they would want to turn back in fear.

Is the magnetic field important to **life on Earth**?

The magnetic field of the Earth is very important to life on Earth because it helps deflect and reflect harmful cosmic rays and solar wind; if we were fully exposed to these rays and wind, the effects upon us would be devastating. When the magnetic fields reverse, there is little or no magnetic field in existence. In the past, some scientists believed that magnetic reversal might have been the reason for the extinction of the dinosaurs.

COMPASSES

How is a **compass made**?

A compass is a magnetized needle placed on a low-friction pivot point. Sometimes the needle is placed in a container of liquid to dampen the movement of the needle. The magnetic needle aligns itself with the north/south orientation of the Earth's magnetic field, and the person using the magnet can determine what direction they are headed in by looking at the compass needle—on the container holding the compass, there are 360 degree markings indicating how far off the direction is from north.

Does a compass sometimes **point downward** along with pointing north?

For hundreds of years, navigators using compasses noticed that on occasion, the compass needle would try to point downward in addition to pointing north. This phenomenon, which went unexplained for several hundred years, was understood and observed by Robert Norman, com-

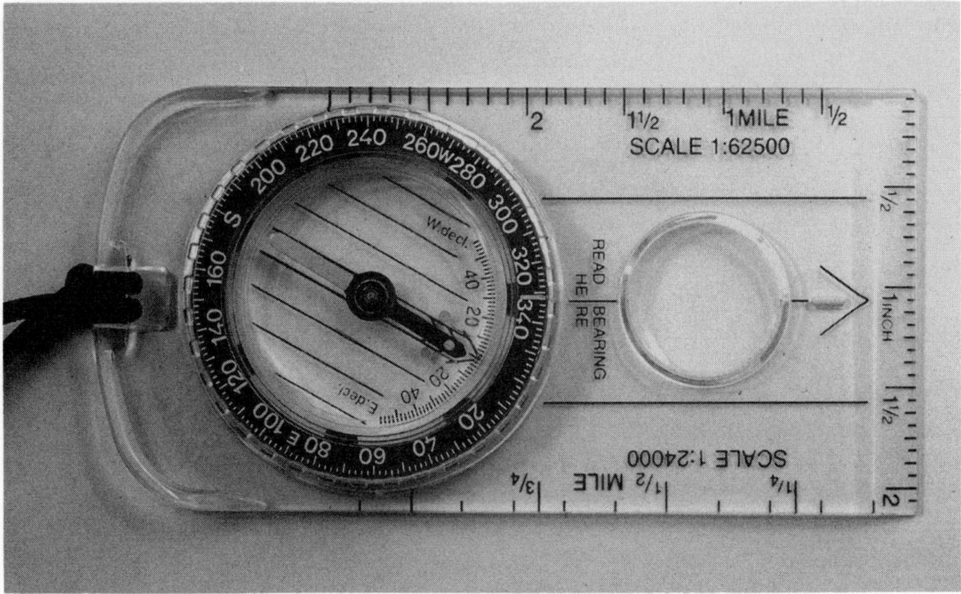

A compass.

pass maker. He found when flying over the poles, one end of the compass would point downward because of its attraction to the pole underneath the plane. He tried to solve this problem by making the compass vertical. With that experiment he discovered the dip needle.

What is a **dip needle** and how is it similar to a compass?

A dip needle is just like a conventional compass, but instead of holding it horizontally, it is held vertically. It is a magnetic needle used for navigational purposes just like a compass, but is used predominantly when traveling around the north and south poles. Instead of measuring horizontal magnetic deflection, the dip needle measures vertical magnetic inclination. When over the equator, the magnetic field of the Earth is parallel to the surface of the Earth. However, the closer one gets to the poles, the less pilots rely on compasses, and the more they rely on dip needles to tell them how close they are to the poles. The closer one gets to a pole, the more vertical the magnetic field becomes, because it's turning into the surface of the Earth. Therefore, when directly over the poles, the dip needle points directly downward.

Do compasses point toward the north and south poles?

Compasses align themselves with the magnetic poles, but not the geographic poles of the Earth. In fact, the geographic poles and the magnetic poles are quite a distance from each other. The magnetic north pole, the pole that affects compasses, is located in northeastern Canada, while the magnetic south pole is situated off the southern coast of Australia. The geographic poles are actually the location of the Earth's axis of rotation.

ELECTROMAGNETISM

How was the **connection** between **electricity and magnetism** discovered?

The close connection between electric current and magnetic fields was discovered quite by accident, and proved to be quite embarrassing. In 1819, Danish physics professor Hans Christian Oersted gave a lecture intended to prove that there was no connection between electricity and magnetism. In the middle of his presentation, he conducted a demonstration in which he placed a compass next to a current-carrying wire. To his dismay, when he picked up the compass and held it above the wire, the compass needle shifted in an east-west direction, indicating that the wire generated its own magnetic field.

How did **Oersted's discovery** affect science?

The fact that moving charge in a wire could create a magnetic field created a great deal of excitement and enthusiasm in the scientific community. A French physicist and mathematician, Andre Marie Ampere, performed variations of Oersted's demonstration and helped pave the way for electromagnetism. Michael Faraday found that by placing a magnet near a wire, the magnet caused charge to flow. Faraday's discovery proved that Oersted's findings were reversible. Not only could electricity create a magnetic field, but a magnetic field could create electrical current.

What connection does the relationship of electricity and magnetism have with the **electromagnetic spectrum**?

The electromagnetic spectrum, as discussed in the WAVES and LIGHT chapters, is dependent on the discoveries of Oersted, Ampere, and Faraday. Electromagnetic waves (including radio, microwaves, visible light, X-rays, and others) are created by the vibrations of electric charge. The vibrating charge creates an oscillating magnetic field, which in turn generates the electrical field. These vibrations propagate outward in the form of two perpendicular, transverse waves, one electric and one magnetic, together called an electromagnetic wave.

ELECTROMAGNETICS AND TECHNOLOGY

Why are **electromagnets,** the kind that pick up junk cars, so strong?

An electromagnet is an iron core (the basis of a magnet) wrapped up in electrical, current-carrying wire. When the current is turned on in the electromagnet, it creates a strong magnetic field, which is made even stronger by the iron core. The incredible strength of such electromagnets allows people to more easily move large metal objects, such as steel cars, from one location to another.

How do **televisions** employ electromagnetic principles?

Electrons fired onto the screen form the images that we watch on the television. In order to create a recognizable image on the screen, each electron needs to be directed to a particular location on the screen. To accomplish this, electromagnets are used deflect the electrons up, down, and side to side within the picture tube. These electromagnets, which are simply coils of copper wire, have varying amounts of current passing through them, and, in effect, vary the amount of magnetic deflection that the moving electrons experience. The combination of thousands of deflected electrons creates the picture on the television screen.

Do speakers use **electromagnets**?

Toward the rear section of a speaker is a permanent magnet that covers a coil of wire. The magnet, attached to the cone (the black paper-like material that vibrates) of the speaker, moves back and forth, depending upon the direction and intensity of the electric current passing through the wire in the electromagnet. It is the current, which pushes and pulls on the magnet and cone, that produces sound by compressing and rarefacting the air molecules around the speaker. (Refer to the SOUND chapter for more information.)

How do **metal detectors** work?

Built into the frame of conventional metal detectors are coils of wire, called solenoids, which carry a current. Whenever metal passes close to these coils, the magnetic properties of the metal change the current in the coils of wire, triggering an alarm.

How do **traffic lights** at car intersections know when a vehicle is present?

Many traffic lights are triggered to change by the approach of a car. The principle is similar to the metal detector, in that the there are coils of current-carrying wire just below the road where the vehicles stop at the intersection. When a large enough amount of metal passes over the coil, it induces a change in current, which causes the light to change.

What are **MAGLEV trains**?

MAGLEV, or magnetically levitated, trains are different from conventional trains in that they use electromagnetic forces to lift the cars off the track and propel them along thin magnetic tracks, which can accelerate trains up to 500 kilometers per hour (300 miles per hour). Although the United States does not currently use MAGLEV technology, Germany and Japan have conducted a great deal of research into the field.

What are the two main forms of **MAGLEV transportation**?

The Germans have concentrated on using a system of electromagnets to lift the underside of the train 1.5 centimeters above its guide rail. This system is difficult to keep stable, but has been used successfully in slower commuter trains.

The Japanese have taken a slightly different approach toward MAGLEV technology. Through the use of superconductive electromagnets under the body of the train, currents traveling through coils in the guide rails levitate the train 15 centimeters off the rail. This type of train, known as a low-flying train, can only levitate if travelling at speeds over 100 kilometers per hour. Otherwise, the train rolls on conventional wheels.

What is the difference between a **motor** and a **generator**?

In each device, a magnet and a coil of wire are employed to change one type of energy to another type of energy.

A motor uses electrical energy to create mechanical energy. Electrical devices that use motors send electricity into a wire loop placed inside a magnetic field. The magnetic field causes the wire to rotate, resulting in mechanical energy. The motor can rotate a fan, blow air out a hair dryer, or turn the blades of a blender.

A generator does the opposite of a motor: where a motor changes electrical energy to mechanical energy, a generator changes mechanical energy to electrical energy. Emergency back-up generators are often used when the regular power goes out in buildings. The generator works off of the mechanical energy produced by an fuel-powered engine, turns a coil of wire in a magnetic field, and forces electrons to flow.

BIOMAGNETS

Can magnets help **reduce pain**?

"Biomagnets," as they are referred to, are tiny magnets priced between $20 and $500, depending upon the size and strength of the magnet. The manufacturer's claim is that the magnetic field from the magnets causes the positive and negative electrically charged particles in the blood vessels to flow faster, which increases the temperature in the vessels. This makes them expand, which increases their surface area and allows the accelerated circulation to increase healing and decrease pain.

Who uses these **biomagnets**?

Although biomagnets do not have approval from the Food and Drug Administration, and many doctors are skeptical, other people, especially athletes, swear by their results. Many golfers and baseball and football players use the magnets to help reduce back pain and other aches associated with their sports. In fact, many people have purchased magnets the size of mattresses. Although there are many who swear by the healing power of magnets, serious independent scientific study has not been conducted to prove or disprove the claims.

VAN ALLEN BELTS

What are the **Van Allen belts**?

Within the magnetic field of the Earth, there are two regions where electrons and protons from solar wind and cosmic rays are essentially trapped between the magnetic field lines and the Earth's atmosphere. They are trapped because the charged particles sneak their way through the magnetic field, and have nowhere to move but within the doughnut-shaped regions called the Van Allen belts. The belts are concentrated around the equator and become thinner as they approach the poles. The two belts are located at 3,200 kilometers and 16,000 kilometers above the surface of the Earth.

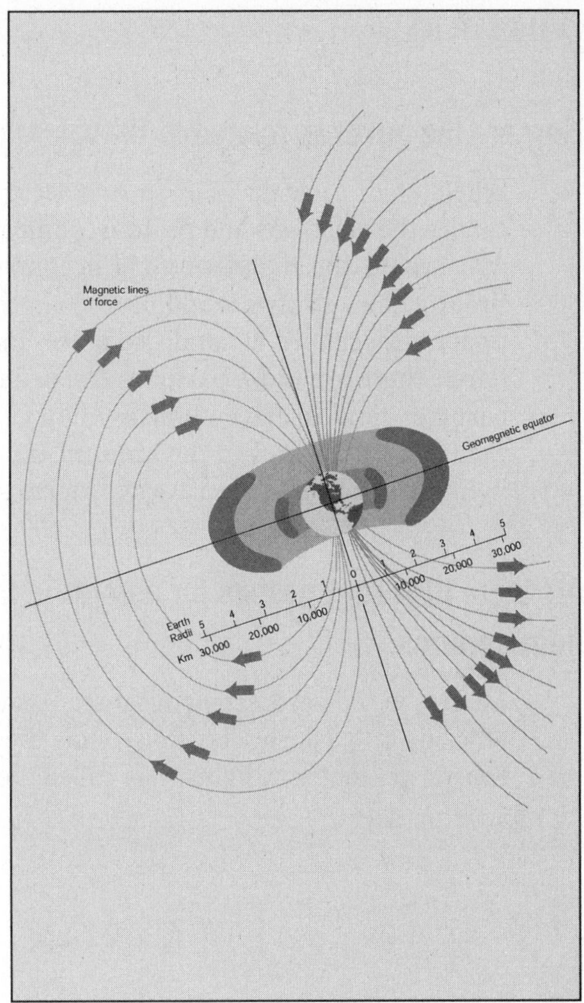

Diagram of the Earth's magnetic lines of force, and the Van Allen belts, two bands of radiation that surround the Earth as a result of its magnetic field.

Why aren't the Van Allen belts present around the **north and south poles**?

At the equator, the magnetic field is parallel to the ground and the electrons and protons from the solar wind can become trapped between the field lines and the atmosphere. However, at the poles, the magnetic field lines are perpendicular to the surface, and are therefore unable to trap the electrons and protons. Particles from the Van Allen belts that are able to reach the polar regions penetrate the atmosphere and produce natural light shows call auroras.

AURORAS

How are the **auroras** related to the Van Allen belts?

When major solar flares or storms occur on the sun, solar wind, which carries the electrons and protons, bombards the Earth's magnetic field. As a result, the magnetic field becomes slightly compressed and, at times, allows electrons and protons in the Van Allen belts to enter the upper atmosphere around the poles. Usually during periods of solar flares, there are more particles trapped from the solar wind and those particles tend to have slightly higher energy. The charged particles excite gas molecules in the atmosphere and create visible light. These lights form beautiful displays of dancing colors in the night sky.

Are there **different names** for auroras in the **northern and southern hemispheres**?

Auroras in the northern hemisphere, known as the "northern lights," are officially called aurora borealis, while the auroras in the southern hemisphere, or "southern lights," are called the aurora australis.

ELECTRONICS

What is **electronics**?

Electronics is the branch of physics that deals with the flow of electrons and other carriers of electric charge. This flow of electric charge is known as electric current, and a closed path through which current travels is called an electric circuit. The study of electronics plays an important role in the development and function of modern technology.

What is a **transistor**?

It was the much smaller and cheaper transistor that would take the place of the vacuum tube in most electronic devices. The first transistor

What was considered to be the advent of electronics?

In 1879, an English physicist named William Crookes designed the first primitive vacuum tube, which consisted of a low-pressure gas in a glass tube. Inside the glass tube were two electrodes that would produce a glow caused by electrons. Crookes discovered that the electrons produced by the tube could be moved up, down, and from side to side by magnetic fields around the tube. It was this tube, later called the cathode ray tube, that would lead to the development of radio, television, and computers.

was invented in 1947 by three American electrical engineers. One of the most important components of any electrical circuit, a transistor controls the current flowing through a particular region of a circuit, allowing the transistor to function as an amplifier. When connected to other transistors, they can be used to help store information in computers, calculators, and other electronic devices.

What is a **semiconductor**?

A semiconductor is a material that can act like an insulator, but also can be made to act like a conductor through the application of a few volts of electricity (generally less than 10 volts). The behavior of a semiconductor is controlled by the application of an easily attained voltage. Semiconductors such as germanium and silicon have only a few free electrons that can flow freely in the material; these materials are used in electronic components to vary the strength and flow of electrons, and are often found in transistors.

What is a **microchip**?

A microchip, the heart of computers and other electronic devices, contains thousands and sometimes millions of miniature transistors and other electrical devices. The microchip, or "integrated circuit," was developed in 1958 by two American electrical engineers named Jack Kilby and Robert Noyce. Since the advent of these chips, originally called monolithic integrated circuits, they have become faster, smaller, and cheaper every few weeks. In fact, the microchips found in calculators, cell phones, and computers today will be considered slow and outdated in just a few months. It is the constant upgrading and developments of new and faster chips that make it impossible for consumers to keep up with technology.

COMPUTERS

I have one on my desk...but what *is* a **computer,** really?

The digital computer is a programmable electronic device that processes numbers and words accurately and at enormous speed. It comes in a variety of shapes and sizes, ranging from the familiar desktop microcomputer to the minicomputer, mainframe, and supercomputer. The supercomputer is the most powerful in this hierarchy, and is used by such organizations as NASA to process upwards of 100 million instructions per second. The impact of the digital computer on society has been tremendous; in its various forms, it is used to run everything from spacecraft to factories, health care systems to telecommunications, banks to household budgets.

What would have been the **first automated computer**?

By 1833, Charles Babbage had already started to work on an improved version of his calculating machine—the analytical engine, an automated programmable machine that could perform all types of arithmetic functions. The analytical engine had all the essential parts of the modern computer: an input device, a memory, a central processing unit, and a printer. For input and programming, Babbage used punched cards, an idea borrowed from Joseph Jacquard, who had used them in his revolutionary weaving loom in 1801.

Where did computers come from?

The story of how the digital computer evolved is largely the story of an unending search for labor-saving devices. Its roots go back beyond the calculating machines of the 1600s to the pebbles (in Latin, *calculi*) that the merchants of Rome used for counting, to the abacus of the fifth century B.C. Although none of these early devices were automatic, they were useful in a world where mathematical calculations, laboriously performed by human beings, were riddled with human error.

By the early 1800s, with the Industrial Revolution well underway, errors in mathematical data had assumed new importance; faulty navigational tables, for example, were the cause of frequent shipwrecks. Such errors were a source of irritation to Charles Babbage, a brilliant young English mathematician. Convinced that a machine could do mathematical calculations faster and more accurately than humans, Babbage in 1822 produced a small working model of his difference engine. The difference engine's arithmetic functioning was limited, but it could compile and print mathematical tables with no more human intervention needed than a hand to turn the handles at the top of the model. Although the British government was impressed enough to invest £17,000 in the construction of a full-scale difference engine, it was never built; the project came to a halt in 1833 in a dispute over payments between Babbage and his workmen.

Although the analytical engine has gone down in history as the prototype of the modern computer, a full-scale version was never built. Among the deterrents were lack of funding and a technology that lagged well behind Babbage's vision. Even if the analytical engine had been built, it would have been powered by a steam engine, and given its purely mechanical components, its computing speed would not have been great. Less than twenty years after Babbage's death in 1871, an American by the name of Herman Hollerith was able to make use of a new technology—electricity—when he submitted to the United States government a plan for a machine that could compute census data. Hollerith's electromechanical device tabulated the results of the 1890 U.S. census in less than six weeks, something of an improvement over the seven years it had taken to tabulate the results of the 1880 census. Hollerith went on to found the company that ultimately emerged as IBM.

What is considered the **first electronic computer**?

World War II was the motivation for the next significant stage in the evolution of the digital computer. Out of it came the Colossus, a special-purpose electronic computer built by the British to decipher German codes; the Mark I, a gigantic electromechanical device constructed at Harvard University under the direction of Howard Aiken; and the ENIAC, another huge machine, but one that was fully electronic and thus much faster that the Mark I. Built at the University of Pennsylvania under the direction of John Mauchly and J. Presper Eckert, the ENIAC cost $400,000 and operated on some 18,000 vacuum tubes. If its electronic components had been laid side by side two inches apart, they would have covered a football field.

Who improved upon the **ENIAC**?

The ENIAC was a general purpose computer in theory, but to switch from one program to another meant that a part of the machine had to be disassembled and rewired. To circumvent this tedious process, John von Neumann, a Hungarian-born American mathematician, proposed the concept of the stored program—that is, coding the program in the same way as the stored data and keeping it in the computer for as long as needed. The computer could then be instructed to change programs, and the programs themselves could even be written to interact with each other. For coding, Neumann proposed using the binary numbering system—0 and 1—rather than the 0 to 9 of the decimal system. Because 0 and 1 correspond to the on or off states of electric current, computer design was greatly simplified.

Neumann's concepts were incorporated in the British-built EDSAC and the University of Pennsylvania's EDVAC in 1949, and in the UNIVAC and other first-generation computers that followed in the 1950s. All these machines were large, plodding dinosaurs by today's standards. Since then, advances in programming languages and electronics (such as the transistor, the integrated circuit, and the microprocessor) have led to computing power in the forms we know it today, ranging from the supercomputer to far more compact models.

How **fast** are **microprocessors** today?

Between the time this "handy answer" was written to the time you read it, the average speed of a new computer's microprocessor will have doubled. The general trend in microprocessor technology has seen the speed of microprocessors double every eighteen months; the technology has continually allowed more and more tiny transistors to fit on microprocessors, and the materials and techniques used to create these integrated circuits continually improve as well. It is estimated that today's microprocessors are approximately 100,000 times faster than the first microprocessor chip developed in the late 1950s.

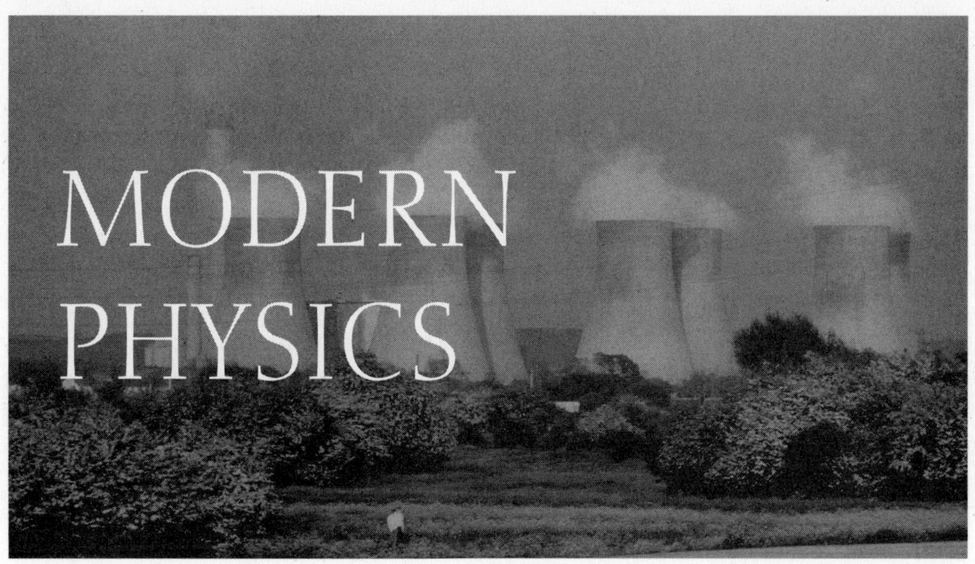

MODERN PHYSICS

THE BASIC ELEMENTS OF MATTER

What is **matter**?

Matter is anything that takes up space and has mass (or weight, which is the influence of gravity on mass). It is distinguished from energy, which causes objects to move or change, but which has no volume or mass of its own. Matter and energy interact, and under certain circumstances behave similarly, but for the most part remain separate phenomena. They are, however, inter-convertible according to Einstein's equation $E = mc^2$, where E is the amount of energy that is equivalent to an amount of mass m, and c is a constant, the speed of light in a vacuum.

In 1804, the English scientist John Dalton formulated the atomic theory, which set out some fundamental characteristics of matter, and which is still used today. According to this theory, matter is composed of extremely small particles called atoms, which can be neither created nor destroyed. Atoms can, however, attach themselves (bond) to each other in various arrangements to form molecules. A material composed entirely of atoms of one type is an element, and different elements are made of different atoms. A material composed entirely of molecules of one type is a compound, and different compounds are made of different molecules. Pure elements and pure compounds are often referred to collectively as pure substances, as opposed to a mixture in which atoms or molecules of more than one type are jumbled together in no particular arrangement.

Subatomic Particles

What are **atoms** made of?

Although it was initially thought that atoms were indivisible, it was later found that they were composed of three principle particles: positively charged protons and neutral neutrons, which contain nearly all of the mass and comprise the nucleus of an atom, and the negatively charged electrons, which have very little mass and reside outside the nucleus.

What are **subatomic particles**?

Subatomic particles are particles that are smaller than an atom. Historically, subatomic particles were considered to be electrons, protons, and neutrons. However, the definition of subatomic particles has now been expanded to include elementary particles and all the particles they can be combined to make.

Elementary particles are particles that cannot be divided up into smaller particles. There are two kinds of elementary particles. One kind of elementary particles make up matter. Examples of these particles include electrons and quarks (which make up protons and neutrons). Baryons and mesons are combinations of quarks and are considered subatomic particles. The most famous baryons are protons and neutrons. The other elementary particles are mediators of the fundamental forces. These mediator particles enable the matter particles to interact with each other. To illustrate this idea, let's say two boys want to play catch. The boys represent the matter particles, and playing catch represents the fundamental force. In this case, the ball would represent the mediator particle.

What is the difference between **hadrons** and **leptons**?

In recent years it seems as though more and more physicists find new and smaller subatomic particles. How do physicists, much less other people, keep track of all these particles? In their endless search to classify, order, and simplify things, physicists have created "families" for subatomic particles.

The lepton family, derived from the Greek word for lightweight, contains particles that interact with each other through the weak force. These particles, of which there are six in all, are responsible for moving particles around and keeping them together. The members of the lepton family are the electron, electron neutrino, tau, muon, muon neutrino, and tau neutrino.

The hadron family, named after the Greek word for thick, interacts with each other through the strong and weak nuclear force, and is actually divided further into two sub-families, the mesons and baryons. Hadrons such as protons, neutrons, and pions, are known for interacting quite strongly with each other and form the building blocks of all matter.

PARTICLE ACCELERATION

How have physicists been able to uncover **new subatomic particles**?

Particle accelerators, such as CERN, in Switzerland, and Fermilab, in Illinois, are designed to accelerate particles such as electrons, protons, and other subatomic particles close to the speed of light so they can collide with other particles. After such energetic collisions, physicists study the aftermath or "particle showers" that often reveal smaller and smaller particles. It is through analyzing such collisions that scientists have been able to find new subatomic particles.

Since the first particle accelerator was designed in the 1920s by Robert Van de Graaff, it has undergone many changes and improvements. The bigger and more powerful an accelerator is, the more it can reveal about the subatomic world. In fact, until recently (construction was halted in 1994 after costs rose well over budget), the United States was constructing the Superconducting Super Collider in Waxahachie, Texas, a 54-mile long accelerator called a storage ring collider. To date, however, some of the most powerful accelerators are found in CERN, which is building a 27-kilometer (16.7 miles) "Large Hadron Collider," and at Fermilab's Tevatron, where the elusive top quark was discovered.

How are particles accelerated to a speed that is close to the speed of light?

Accelerators use massive magnets to guide and accelerate particles to speeds unimaginable to the average person. Some accelerators actually employ the use of transmitters, which boost the energy of the particle until it achieves the desired speed before colliding in the detector.

What is a **linear accelerator**?

A linear accelerator is a particle accelerator that is laid out in a straight line. A proton is fired into the accelerator, and by alternating the charges in the tubes that the electron or protons travels through, it is accelerated along the 3.3-kilometer (2 miles) track until the desired energy is obtained. Upon its collision, the paths and particles emitted from the collision are analyzed and studied by particle physicists. The longest and most powerful linear accelerator is found at Stanford University in California. The accelerator at Stanford is specifically designed to accelerate electrons up to extremely high energies.

What is the difference between a **synchrotron** and a **linear accelerator**?

A synchrotron, otherwise known as a frequency modulated cyclotron, serves the same purpose as a linear accelerator, but instead of accelerating the particles in a straight line, the synchrotron uses massive magnets to repeatedly accelerate the particles around a circle. When the velocity of the particle is fast enough, physicists send the particles into targets where they can observe the destruction and perhaps catch a glimpse of some new, fundamental particle. An example of a synchrotron is the Tevatron at the Fermi National Accelerator Laboratory on the outskirts of Chicago.

What happened to the construction of the **super collider** in Waxahachie, Texas?

The Superconducting Super Collider (SSC), begun in 1990 and intended for completion in 1999, was canceled by the United States government in 1994. Although nearly 20% of the collider was complete, Congress, under pressure to reduce the growing national deficit, stopped construction on the proposed $8 billion super collider after projected costs escalated to $10 billion.

If the **Superconducting Super Collider** had been completed, what could it have done for physics?

The choice to begin and then end construction of the SSC in Texas was a hot debate in the late 1980s and '90s. Regardless of the debate, the SSC would have been far superior to any particle accelerator of its day. The accelerator, measuring 87 kilometers (54 miles) in circumference, would be able to accelerate and crash protons together at a rate of 50 million per second. Other improvements over other accelerators would be a magnetic field 50% stronger than Tevatron at Fermilab. Such major improvements would have allowed physicists to pursue major breakthroughs in particle physics, create simulations that would mimic what occurred milliseconds after the big bang, and allow more in-depth studies into the symmetry of the Grand Unified Theory. (See the DEEP THEORIES chapter for discussion on the Grand Unified Theory.)

QUARKS

What is a **quark**?

The quark is considered the building block of everything. As late as the 1950s, many physicists felt that the fundamental building blocks of nature were the proton, neutron, and electron. That ended in 1964 when two American physicists, working independently, developed a theory that suggested there was something even smaller than the proton

How are quarks classified?

The six quarks are grouped into three generations of two quarks each: "up and down," "strange and charm," and "top and bottom." In addition, each quark carries another property called "color," which comes in red, green, blue, and all the anticolors. The name "color" doesn't have anything to do with visual perception; it is just a name chosen by someone with a whimsical sense of humor. Quarks have fractional charge—plus or minus 2/3 or plus or minus 1/3. Three quarks together is a baryon, and two quarks together is a meson.

and neutron. These physicists, Murray Gell-Mann and George Zweig, found that the most basic particle in the universe was the quark, subdivided into three types: up, down, and strange. They determined that the three "quarks" (the name was taken from a James Joyce novel) were the basic building blocks of the proton in the nucleus of an atom.

Later, work in subatomic particle theories prompted particle physicists to search for the existence of a fourth, fifth, and sixth quark. The fourth and fifth particles were proved to exist after colliding subatomic particles in particle accelerators and watching the aftermath of such energetic collisions. By the late 1970s, physicists felt that there were six fundamental quarks, but the sixth—the "top quark"—had yet to be discovered.

What is the **top quark**?

By 1977, the first five quarks had been discovered. Nearly two decades later, a group of physicists at the Fermi National Accelerator Laboratory in Illinois conducted a series of collisions in their Tevatron accelerator, the most powerful particle accelerator in the world, which accelerated protons and antiprotons toward each other at speeds close to the speed of light. Scientists at Fermilab knew they had their work cut out for them, for the top quark—if they could find it—would only be seen in about one in one billion collisions and last for less than a billionth of a second.

Finally, on March 2, 1995, the top quark was found. It took scientists at Fermilab almost a year to find, was much heavier than any other subatomic particle. The discovery of the top quark, probably one of the most significant recent discoveries in physics, will give scientists more clues to the underlying foundations of matter throughout the universe.

NEUTRINOS

See also: DEEP THEORIES chapter, "THE EXPANDING UNIVERSE"

What are **neutrinos**?

Neutrinos are theoretical particles that are emitted during radioactive beta decay. The beta decay, which is the product of a neutron decaying into a proton and an electron, needed a neutrino to balance the equation. For many years scientists have wondered exactly what this theoretical particle called the neutrino really was. Neutrinos have no electrical charge and, for all intents and purposes, could not be detected until huge underground tanks were built to detect the elusive neutrino.

Why build **underground tanks** to pick up neutrinos?

Neutrinos, the byproduct of beta decay, have no charge and an unknown mass. In fact, it is extremely difficult to prove their existence. It is thought that millions of neutrinos pass through our bodies every day, but they do not harm us in any way. The reason why detection tanks have been built underground is so that neutrinos, which can pass through the ground, will not be confused with cosmic rays, which do not penetrate the ground. The study of neutrinos is rather important, for many physicists feel that up to 90% of the universe consists of neutrinos. Although physicists have not detected the number of neutrinos they had anticipated, they were pleasantly surprised when a ten-second burst of neutrinos bombarded the Earth and the underground tanks when they were emitted from the Supernova 1987-A.

OTHER SUBATOMIC PARTICLES

What are **gluons**?

Gluons are actually similar to what they sound like; they are the subatomic particles that "glue" the quarks together. When quarks are tightly packed together with a proton or a neutron, there are strong repelling forces. If it were not for the gluons holding the nucleus together, the atoms would be liable to fly apart.

What is a **positron**?

A positron, as named by Paul Dirac in 1929, is a subatomic particle that was predicted in mathematical calculations and then later found in 1932. The positron is a mirror image of an electron. It has the same mass, yet has the opposite charge of an electron.

QUANTUM PHYSICS

What is **quantum physics**?

Quantum physics, otherwise known as quantum mechanics, is the theory used to provide an understanding of the behavior of microscopic particles such as electrons and atoms. However, it is more than just a means to calculate, for example, where an electron might be; the quantum theory also introduced an entirely new way of thinking about very small objects that is strangely different from the way we think about macroscopic (that is, larger) objects.

What is antimatter?

Antimatter was predicted in a series of equations derived by Paul Dirac. He was attempting to combine the Theory of Relativity with equations governing the behavior of electrons. In order to make his equations work, he had to predict the existence of a particle that would be similar to the electron, but opposite in charge. This particle, discovered in 1932, was the antimatter equivalent of the electron and called the positron. Other antimatter particles would not be discovered until 1955 when particle accelerators were finally able to confirm the existence of the antineutron and antiproton.

An example of a macroscopic object is a baseball. Whenever we throw a ball into the air, the most exact way to describe the ball's motion is by using "classical mechanics" (or "classical physics"), which predicts the position and velocity of the ball at every instant during its flight. This approach fits our everyday experience, since we are accustomed to seeing a ball move in a very well-defined path.

The problem comes when we try to apply this classical approach to microscopic objects. If an electron were just an exceptionally small ball, its motion would follow a path predicted by classical mechanics. However, experiments have shown that this is not the case, so a new approach to physics was needed to deal with very small particles.

Quantum physics came into existence around the beginning of the twentieth century. It was physicists such as Max Planck, Albert Einstein, and Niels Bohr who were instrumental in establishing the building blocks for quantum physics.

Although quantum physics is relatively new, the theory has been extremely successful in explaining a wide range of phenomena, such as how electrons move in materials, like those that travel through the chips in a personal computer. Quantum mechanics is also used to understand superconductivity, the decay of nuclei, and how lasers work, among many other things. A great number of scientists now use quan-

What did Albert Einstein win a Nobel prize in physics for?

Technically, it was not Einstein's theories of Special and General Relativity that won him a Nobel prize, but rather his work on the photoelectric effect that helped support his quantum theory of light. Einstein believed that quanta of light behaved like a particle. Einstein found that electrons would be emitted from a photosensitive metal surface if the light hitting that surface had enough energy. Einstein further added that it was not the intensity of the light, but the energy of the particular frequency of the light that determined if electrons would be released from the material.

tum mechanics daily in their efforts to better understand the behavior of microscopic parts of the universe. However, the basic ideas of the theory still conflict with our everyday experience and the argument about their meaning continues among the same physicists and chemists who use the quantum theory.

What is **quanta**?

Quanta is the smallest amount of something. Light travels as photons called quanta. It was Max Planck who, around the turn of the century, determined the relationship between the amount of energy that light has and its frequency. Through his mathematical calculations, he found that light's energy is directly proportional to its frequency on the electromagnetic spectrum.

What connection do **light sensors** have with the **photoelectric effect**?

One of the most useful applications of the photoelectric effect has been light sensors. Many light sensors use a photosensitive phototube that receives light on a photosensitive metal and translates the light into electrical impulses in a circuit. Such circuits have been used as safety devices for electric garage door openers, customer counters in stores, light meters in cameras, and as optical soundtracks in reel-to-reel movies.

Einstein felt that light had particle-like characteristics. Can **particles** have **wave-like characteristics**?

Indeed they can. French physicist Louis de Broglie in 1923 found that all particles exhibit wave-like properties. He theorized that when electrons are fired through narrow slits, there would be no way to predict the direction of the electron's path. When enough electrons passed through the slits, a wave-like diffraction image would appear behind the slits. From de Broglie's theories on light's wave-like characteristics, for which he won a Nobel prize, evolved the electron microscope.

What is the **Uncertainty Principle**?

De Broglie realized that the path of a single electron could not be determined. He felt that since traditional physics could not predict the placement of the electron, then quantum mechanics, which relies on probability and randomness, must be used. This principle of randomness used by de Broglie was actually developed and known as the "Heisenberg Uncertainty Principle." Einstein felt extremely uncomfortable with the idea that science must rely on probability, and in response to the uncertainty principle said, "God does not play dice with the Universe."

CHAOS

What does **chaos** mean in physics?

Chaos, by definition, means a lack of order. A famous question posed by meteorologist Edward Lorenz of the Massachusetts Institute of Technology asks, "If a butterfly in China flapped its wings, could it affect the

What are examples of chaotic systems?

Weather is a chaotic system; weather forecasters have difficult jobs, for they attempt to instill order in chaos by predicting the outcome of a weather system. Another, less obvious, chaotic system is our solar system. This means that although the movement of planets in our solar system can be tracked, small variations in the movements of the solar system could cause changes in the next 100 million years. The behavior of chaotic systems, such as the weather and the solar system, can only have predictable results for a limited time.

weather here in North America?" No one knows for sure—and that is the whole point behind chaos. Although we can predict certain events, there comes a point where we can no longer predict what will happen, because even the smallest disturbance combined with other disturbances can have a major affect down the line.

LASERS

What does **stimulated emission** mean?

Whenever an electron jumps from a higher-energy state to a lower-energy state, energy in the form of a photon is emitted from the atom. This stimulated emission of photons occurs when a photon strikes an atom, causing an electron in the atom to decrease its energy level and emit a photon. As a result, two photons are emitted from the atom: the original photon that struck the atom, as well as the photon that was emitted when the energy level of the electron decreased.

Stimulated emission occurs only when a photon is deliberately fired at an atom to produce another photon. The repeated process of emitting photons, all in phase with each other, is necessary for laser light.

What is a **laser**?

The word "laser" is actually an acronym for "light amplification by stimulated emission of radiation." A laser is a device that creates pure, coherent light by reducing the energy levels of electrons, causing them to emit photons (see above question on stimulated emission). It is the repeated process of emitting photons that creates the laser light. The photons, located inside a tube, are stimulated to higher-energy states while reflecting repeatedly back and forth off silvered mirrors. This is called light amplification. Some of the single-frequency, high-energy light escapes out the partially silvered surface on one end of the tube as an extremely narrow beam of light. The light does not spread out, because the only photons allowed to escape from the tube are the photons that move perfectly perpendicular to the partially silvered mirror.

What are some **uses** for **lasers**?

When it was first invented, the laser was called "a solution looking for a problem" because few good applications could be found for it. This is no longer the case; lasers have become extremely useful in almost all aspects of science and technology. In science, physics especially, lasers have been used as trigger mechanisms or switches, measurement tools for time and distance, and in the development of holograms. Lasers are used in industry to drill, cut, and fuse materials and electronics components together. The military has also taken advantage of the laser in guidance, defense, and nuclear weapon systems. It is in the fields of communications and medicine that lasers have had the greatest impact on people, which are the topics of the next two questions.

Can lasers help **cure medical ailments**?

The high-intensity, highly focused light found in the lasers has had a huge impact on the medical profession and the treatments that it provides. Lasers have been used to kill skin cancer cells, fuse sections of the retina, remove birthmarks and moles, and even kill tumors inside the body. New medical laser technologies are always on the forefront of medical technology. For example, scientists and doctors are working on a new method of using lasers to clear blood vessels of plaque and other harmful deposits. Also, lasers are currently being used to correct near-

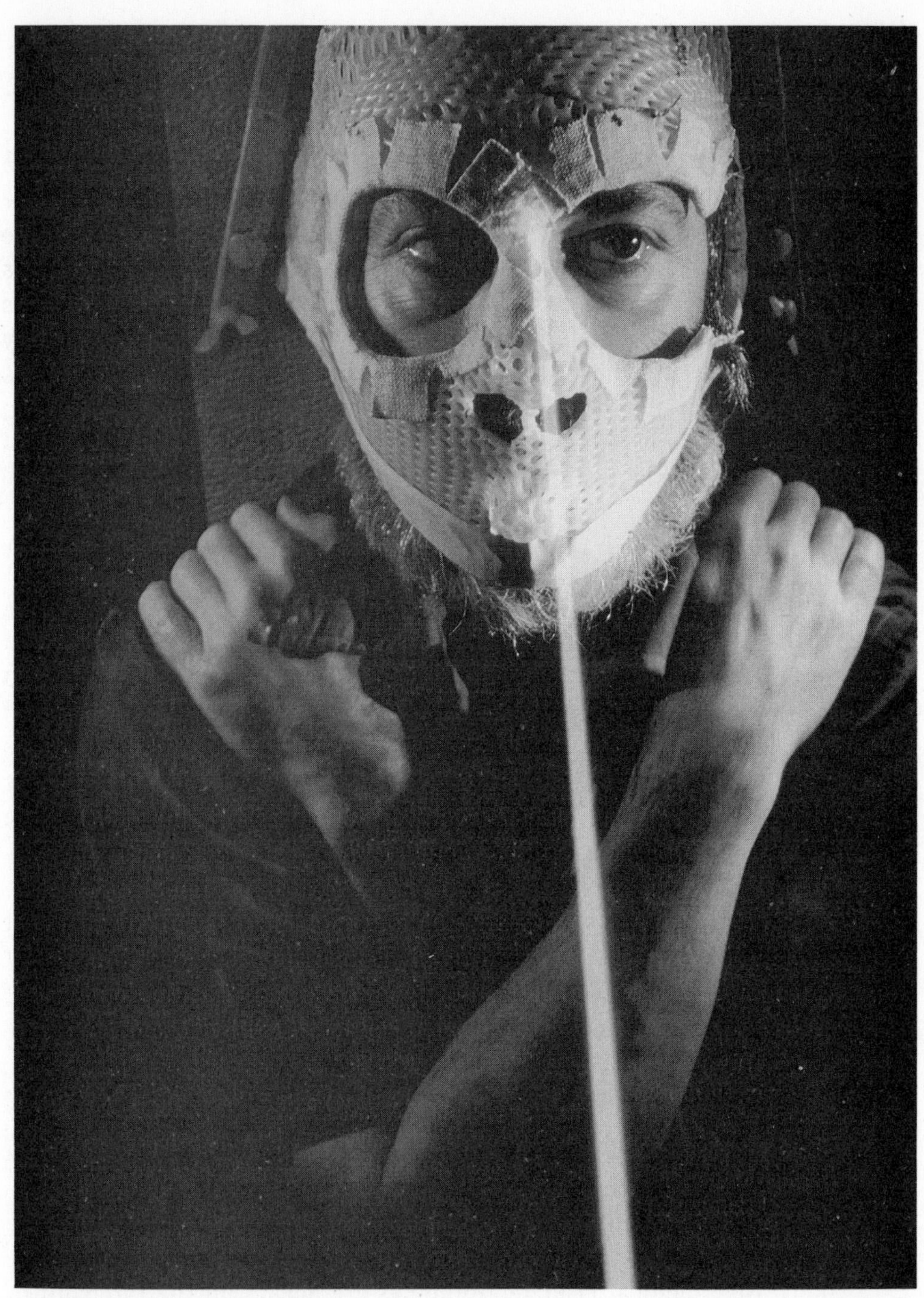

Patient undergoing laser surgery for cancer.

Music CDs use lasers to record and play back music.

and farsightedness. Special argon lasers are being used to shape the cornea of the eye so that the light focuses perfectly on the retina. This, when completely developed and tested by the government, would eliminate the need for most eyeglasses. These are all impressive applications, but this is only the beginning of medical laser technology.

How have lasers been used in **communication**?

Just as lasers have been used extensively in the medical field, lasers have had just as large an impact on the communications field. For example, pulsed laser signals aimed into fiber optic cables have been used to instantaneously transmit thousands upon thousands of telephone and television signals. This is much more than any other form of communication medium could ever transmit. Soon computers will transmit information through fiber optic wires driven by lasers. Such technologies will allow information to flow quickly, but will also prevent the overheating of electrical components that conventional systems experience.

Digital compact discs (CDs) also use lasers to read and record information. Compact disc lasers are aimed at an information layer consisting of bumps

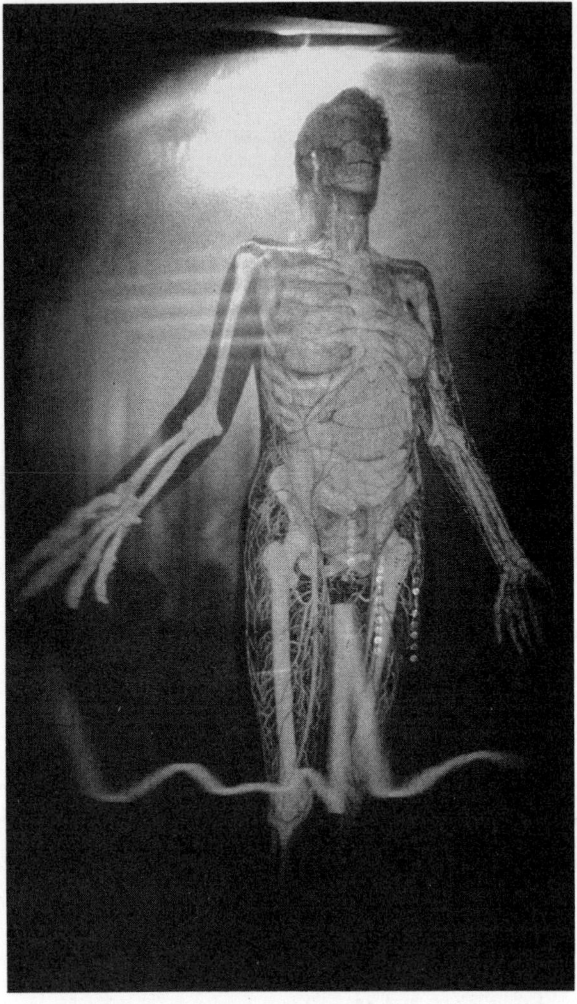

Holograms are created using laser beams.

and valleys, each of which, when combined with approximately 40,000 other bumps and valleys, corresponds to a bit of digital information. The fact that there can only be two possibilities, either a bump, given a value of one, or a valley, given a value of zero, makes the translation of information from a CD flawless. These bits of information, when combined, translate into many different sounds and other types of information.

How do **holograms** use lasers to store information?

Holograms look like three-dimensional images held within the confines of a two-dimensional piece of film. They are often used on credit cards because they are relatively difficult to reproduce and therefore serve as a deterrent to counterfeiters making fake credit cards. Holograms are created by a laser beam that splits up and reflects off different sections of an object and gets recorded on a special type of photographic film that records interfering wave fronts from two or more light sources. When this recording is properly illuminated, a reconstruction of the original wave fronts is formed, and the original image is seen.

RADIOACTIVITY

What **force** keeps the subatomic particles in an atom from **flying apart**?

Although protons and neutrons have mass and can be influenced by gravity, it is not the gravitational force that holds these particles together, but a different force called the "strong nuclear force." This force acts only on subatomic particles, and loses its attractive abilities when the distance is larger than 10^{-15} meters away.

What needs to be done to **break apart a nucleus**?

The amount of energy needed to break the attraction between the proton and neutron in a nucleus is called the binding energy. The stronger and more stable the nucleus, the greater the binding energy needed. Albert Einstein's formula $E = mc^2$ represents the amount of energy that would be needed to break the strong force of a nucleus.

If you take a set of protons and neutrons that are free and weigh them, then assemble that set into a nucleus and weigh it again, you would find that the nucleus is a little lighter than the set of protons and neutrons that were used to assemble it. The difference in mass between the nucleus and its constituents times c^2 (the speed of light squared) is the binding energy of this nucleus. If you wanted to tear apart the nucleus, you would have to do that much work to reduce it to the constituents again.

What is radiation?

Radiation occurs when the nuclei of a radioactive particle, such as uranium or plutonium, disintegrates over time, producing two or more radioactive particles during that disintegration. Radioactive particles can be produced both naturally, as occurs every day in the air around us, and can be artificially produced by humans in nuclear reactors. Many people fear radiation and radioactivity for the harmful ionizing effects it can have on human cells and tissue if overexposed; however, radioactivity can be extremely helpful to humans by helping treat several types of illnesses.

What are some **radioactive particles**?

There are several different types of radioactive particles, all of which can be quite helpful and harmful, depending upon the situation and the exposure levels. The first major type of radioactive particle is the alpha particle. It is essentially a positively charged helium atom consisting of two protons and two neutrons. It is quite difficult for alpha particles to penetrate most materials—in fact, a thin piece of paper is sufficient to block an alpha particle.

Another well-known radioactive particle is emitted after a neutron (which on its own is quite radioactive itself) in the nucleus of an atom decays to a proton, and in the process fires out an electron, called a beta particle. The negatively charged beta particle can travel much greater distances that the alpha, until it also loses its energy after collisions with other atoms and electrons. The beta particle, which is more substantial than the alpha particle, can penetrate paper and other similar materials but can be blocked by a sheet of aluminum.

Finally, gamma rays are the potentially most dangerous of all radioactive rays. Found on the electromagnetic spectrum with an extremely high frequency, the gamma ray is simply a form of non-visible light or photons. Gamma rays are produced when the nucleus of an atom moves from a high- to a low-energy state. Gamma radiation has such high frequency and energy that it can penetrate almost anything. However, the absorbing characteristics of lead provide good protection against gamma rays.

Who **discovered** radioactive decay?

In 1896, Antoine-Henri Becquerel, while conducting experiments with uranium compounds, found that they spontaneously emitted radiation, and that the intensity of the radioactivity depended upon the amount of uranium that was present in the experiment.

What important discovery did **Marie Curie** make regarding radioactivity?

Marie Curie, along with her husband Pierre Curie, was the first to call the breakdown of a nucleus "radioactivity." The couple discovered as many as forty radioactive elements beyond what Becquerel had found. In 1903 Marie and Pierre Curie shared the Nobel prize in physics with Becquerel for their research and breakthroughs in radioactivity.

What **other member** of the **Curie** family won the Nobel prize?

Marie Curie, whose father was a high school physics teacher in Poland, had a daughter who also became a physicist. In 1935, Irène Curie and her husband, Frédéric Joliot-Curie, won the Nobel prize in chemistry for artificially creating radioactive elements.

What is a **half-life**?

The half-life of a process is an indication of how fast that process proceeds—a measure of the rate or rapidity of the process. Specifically, the half-life is the length of time that it takes for a substance involved in a process to diminish to one-half of its initial amount. The faster the process, the less time it will take to use up one-half of the substance, so the shorter the half-life will be.

The half-life of a radioactive element is a means of measuring how long it takes one half the radioactive nuclei to disintegrate. Depending upon

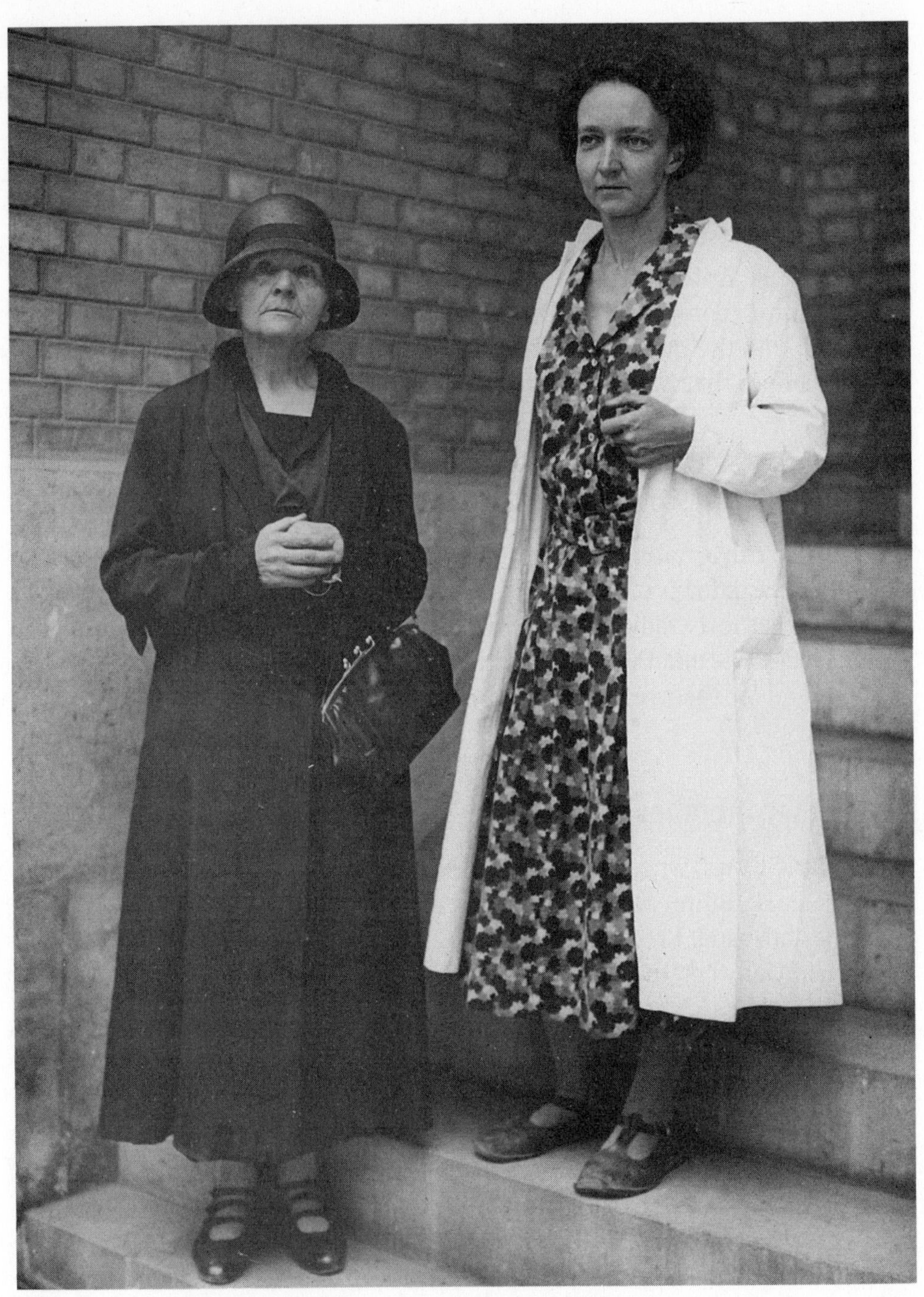

Marie Curie and her daughter, Irène Joliot-Curie, both Nobel prize–winning physicists (see previous page).

How can radioactivity determine the age of something?

Archeologists and anthropologists use the radioactive half-life of the carbon-14 atom to help them date ancient artifacts. By analyzing the number of carbon-14 atoms emitting beta radiation in an archeological sample, and comparing that to the amount of radiation from a new sample of carbon-14, scientists can determine the approximate age of the sample. For example, the half-life of a carbon-14 atom is 5730 years. If a bone sample excavated from an old Native American settlement emitted 25% the beta radiation that a new sample of carbon-14 emits, then the sample would be approximately 11,460 years old.

the particular radioactive material, the time for half the nuclei to decay could be anywhere from a fraction of a second to thousands and even billions of years. For example, the isotope Polonium has a half-life of only 16 thousandths of a second, whereas Indium can exist for 4.41×10^{14} years before half of its nuclei decay.

How is radiation **detected**?

Radiation is detected by the ionization that it produces. The most common tool for detecting radiation is the Geiger counter, which consists of a cylindrical metal tube filled with gas. When reacting to the gas in the chamber, the high-energy, radioactive particle ionizes the gas molecules and emits electrons that the detector will pick up as a radioactive "click" on the Geiger counter. Other instruments that detect radioactive decay are photosensitive scintillation counters, semiconducting detectors, and bubble chambers.

What can **radiation** do to you?

Radiation does damage to you by ionizing material in your cells. If there is a heavily ionized path through the cell wall, the cell may rupture, and

A man using a Geiger counter to detect radiation (see previous page).

if the DNA is ionized the cell can mutate into cancer. On the other hand, a cell may have an ionization path through it, but nothing important in the cell is affected, so the cell sustains little or no damage. Ionization paths can also introduce a lot of free radicals into the cells, which may alter their chemistry. There are many possible effects.

Normal background levels of radiation are not high enough to cause humans any harm. However, overexposure to radiation such as gamma and beta radiation prove quite dangerous. Gamma radiation is considered more dangerous than beta for it is more difficult to stop and can affect every organ and tissue in the human body. People who work in areas where the radiation levels are higher than normal run the risk of getting cancer, and if pregnant, have a much higher risk of birth defects.

Radiation can also help people. Some cancer patients undergo radiation treatments for their cancer. Radiation, although harmful to the patient, can kill certain cancer cells and even force the cancer into remission. Cancer cells are more likely to die when exposed to radiation than are normal cells. So radiation treatment amounts to exposing the patient to a nearly lethal dose of radiation—one that won't quite kill the normal cells but is sufficient to kill the cancer. One nice thing about radiation from non-neutral particles, for instance protons, is that most the damage from radiation is done very near the end of the path, just before the particle stops. You can shoot a tumor with a radiation beam, which does very little damage on the path in, but a great deal of damage on the tumor site. You can shoot the beam in along various paths for different treatments so that the paths all intersect at the tumor; the tumor is damaged for every treatment, but the cells on the incoming paths only see radiation for one or two treatments. The rest of the treatments are along different paths.

NUCLEAR PHYSICS

NUCLEAR REACTIONS

What is a **nuclear reaction**?

Simply put, nuclear reactions are interactions involving collisions between atomic nuclei. Such reactions are the source of the energy generated by nuclear power plants and delivered by nuclear weapons. They are also the source of the energy that enables stars to shine and so are at the heart of stellar astrophysics.

What are the **two types** of nuclear reactions?

Fission and fusion are the two methods of creating a nuclear reaction. Fission creates a nuclear reaction by splitting a heavy nucleus into two lighter nuclei, which creates a chain reaction of similar reactions. Fusion, on the other hand, is the opposite of nuclear fission. Whereas fission separates an atom, fusion produces energy when a light nucleus is fused together with another light nucleus to form one nucleus. After the fusion process, the mass of the particle is smaller because some of that mass has been converted into nuclear energy. As of yet, all nuclear power is produced through fission, while scientists are still working on efficient methods of sustaining fusion reactions.

FISSION

Who was **Enrico Fermi** and what was his contribution to nuclear physics?

Although the first sign of a nuclear fission reaction was observed by Otto Hahn and Lise Meitner in Berlin, Germany, the first controlled nuclear reaction did not occur until Enrico Fermi conducted some experiments at the University of Chicago in 1942. It was on December 2 of that same year that Fermi, an Italian physicist who immigrated to the United States before World War II, was first able to sustain a fission reaction. Fermi (for whom the National Accelerator Laboratory, outside of Chicago, is named) worked in Los Alamos, New Mexico, figuring out how nuclear power could be used against the Japanese. Although he helped work on fission, Fermi—along with many other scientists—felt that nuclear weapons should not be used in war.

What are some **advantages** to nuclear fission?

After World War II, many people felt that fission would be a new and abundant source of electrical power. Indeed, as little as one ounce of Uranium-235 could produce more energy than 100 tons of coal could. In addition, just one gram of U-235 could create 18,000 kilowatt-hours of heat. Nuclear fission could also eliminate many of the pollutants otherwise generated from coal, gas, oil, and wood. Finally, it is not only the amount of energy produced that is so amazing, but the fact that such a reaction could remain easily sustainable through its chain reaction without using more energy to keep the reaction going.

Enrico Fermi.

What are some **drawbacks** to nuclear fission?

Many people fear nuclear fission power. Reasons for such fear might be the possibility of exposure to low levels of radioactivity and radioactive wastes. With the stringent safety precautions that are in place today, such fears are not warranted. Other major fears stem from the memories of events such as the nuclear meltdowns at Three Mile Island and Chernobyl. Although the Three Mile Island incident was relatively contained, the Chernobyl nuclear accident resulted in massive amounts of radiation leaking out into the surrounding region and the world.

What is a **chain reaction**?

A chain reaction is a repeating series of events that continue without stopping. Chain reactions in fission reactions are initiated when a neutron strikes a U-235 atom, splitting that atom into two other atoms and three other neutrons. These neutrons strike another U-235 atom and the event occurs over and over again. The chain reaction will only stop when there is not enough uranium left to split.

Is it difficult to find Uranium-235?

Although it is not very difficult for countries and organizations to find uranium, it is specifically the U-235 that is difficult to separate from the more abundant U-238. In order to do so, tons of U-238 is mined, crushed, and then chemically separated into pure uranium, of which less than 1% is U-235.

What does it mean to reach **critical mass**?

In order for a chain reaction to occur, one fission event triggers one or more other fission events. If an average of one neutron from a fission reaction causes another fission reaction, then it has reached its critical mass and the reaction will continue. If, however, the average number of neutrons from the fission event produces, on average, less than one other fission reaction, then the chain reaction has become subcritical and will end shortly. If more than one neutron from each reaction sustains the fission process, the reaction has reached a supercritical stage.

What is the difference between **heavy water** and **light water** reactors?

Countries and nuclear power plants that have access to enriched U-235 are able to use pure water as a coolant for the reaction. Such a reactor is called a light water reactor (LWR). Plants in countries that do not have access to enriched samples of U-235 have had to use heavy water (deuterium oxide) as a coolant. These were called natural uranium reactors for they did not have the ability to create an enriched version of the uranium.

Can other elements **besides uranium** be used in nuclear fission reactions?

There is another element that can be used to create nuclear fission in power plants and nuclear weapons. Plutonium, derived from the beta

An anti-nuclear demonstrator wears a costume of "Goodman Death" in a demonstration commemorating the 10th anniversary of the Chernobyl power plant explosion.

decay of a Uranium-238 atom, can, when struck by a neutron, create a nuclear fissionable chain reaction. Creating a plutonium atom from uranium-238 to form a nuclear reaction is called breeding, and takes place inside a special breeder reactor. However, unlike uranium, plutonium is an extremely dangerous and toxic radioactive element that must be treated with great care.

What made the **Chernobyl** nuclear accident the most devastating, non-wartime nuclear accident of all time?

On the morning of April 26, 1986, problems with the water-cooling system in one of the nuclear reactors of the Chernobyl nuclear power plant in the Ukrainian republic of the former U.S.S.R. led to an explosion and fire, causing a massive meltdown in the core of one of the reactors. Instead of having safeguards that would shut down power if a nuclear meltdown occurred, most Soviet nuclear reactors were instead programmed to boost power, thinking that it needed to increase production. Unlike the Three Mile Island incident, the Chernobyl nuclear

A Pennsylvania state policeman and plant security guards stand outside the closed front gate to the Metropolitan Edison Nuclear Power Plant on Three Mile Island after the plant was shut down following an accident on March 23, 1979.

power plant's reactor was not surrounded by thick concrete. Instead, the massive amount of steam caused by the overheating nuclear core blew the 1000-ton steel roof off the reactor, sending tens of millions of curies (a unit of radioactivity) into the immediate area and atmosphere.

In the days and weeks that followed the explosion, wind currents throughout Europe's atmosphere carried the radioactivity in clouds, dropping contaminated rain on many countries throughout western and central Europe and Scandinavia, as well as over the Soviet Union. The immediate death toll from the explosion and exposure to high amounts of radiation was 31, while hundreds of thousands of people in the Soviet Union and Europe were exposed to abnormally high amounts of radiation from nuclear fallout in the days following the accident.

What happened at **Three Mile Island** in Pennsylvania?

The 1979 accident at the Three Mile Island nuclear power plant near Harrisburg, Pennsylvania, occurred as a result of a coolant pump gone bad—there was no coolant to keep the control rods (which are used to keep the fission reactions under control) in the reactor from melting. Although meltdown did occur, huge concrete walls blocked most of the radiation that would have otherwise escaped into the air.

What does **E = mc² ** mean?

One of Albert Einstein's most famous discoveries was that energy and mass are the same thing, just in different forms. What this means is that this book resting in your hands is actually a form of energy in storage, called rest energy or mass. If the book, which is now mass, could be completely changed into energy, it would yield an enormous amount of energy.

In his Special Theory of Relativity, Einstein produced the mass-energy equivalence formula known as $E = mc^2$, where E is the energy, m is the mass of the object, and c^2 is the speed of light squared. This formula allows physicists to calculate how much mass is needed to create a certain amount of energy (if all the mass could be completely converted to energy.)

This concept can be described using an automobile as an example. The fuel that a car uses has mass. That fuel goes through a chemical reaction that changes some of the fuel into exhaust, which also has mass. However, some energy is present during the combustion of the fuel—energy that did not appear to be present before the fuel combusted. That energy actually came from the fuel. A tiny amount of the fuel's mass was converted into energy. In fact, the amount of exhaust mass is now slightly less than the mass of the original fuel, because some of the fuel's mass was converted into usable energy. A similar concept is used in nuclear reactions where mass is manipulated (fused or split apart) to create energy.

FUSION

What type of energy is used to **light up the sun**?

The energy that lights up our sun as well as all other stars is produced by nuclear fusion. In order to create a fusion reaction, two atoms must join together, and in the process, release a huge amount of energy. Whenever two lighter atoms join together to form a larger, more massive atom, the resultant atom has less mass than the original two atoms. Again, just as in fission, some of the mass has been converted to energy.

What does the **future** hold for fusion?

Fusion is probably the only method of producing energy that could serve as a long-term energy source for our world. Fusion power plants would be perfectly safe, free from radioactive waste, and non-polluting. The fuel for fusion would be extremely easy to obtain for it is available throughout the world. Finally, the power unleashed by a fusion reaction would be unparalleled by any other form of energy. For example, the resultant energy from a deuterium-tritium fusion reaction would be over 400 times the energy that would have to be put into the system to get the reaction to occur.

How could fusion be **generated** by humans?

To generate a fusion reaction, the two particles that will be fused must be stripped of their electrons and moved extremely fast. In order to prevent repulsion of the two positive nuclei, the temperature of the particles is raised to many times that of the sun. In fact, the temperature gets so hot that instead of remaining in a gaseous form, the particles become a different phase of matter, called plasma. Upon fusion, the nuclei release huge amounts of energy and, as a result, have less mass than what they started out with. Today, one of the major problems with fusion is containing the high-temperature plasma in something that will not melt or allow it to escape.

What are some methods of **containing the plasma** during thermonuclear reactions?

There are three ways of containing plasma during a fusion reaction. One of the methods uses a strong magnetic field to protect the materials inside the reactor and prevent the contents from escaping. The second

The tokamak fusion test reactor at the Princeton Plasma Physics Laboratory in Plainsboro, New Jersey.

method of confining plasma is through inertial confinement, which keeps the plasma together by firing multiple laser beams, such as found in the Nova Laser, at targets within the reaction chamber. The final method of containing plasma is by using gravity, but the only "reactors" that are able to do this are not man made; only the sun and stars have thus far been able to contain plasma this way.

What is the **tokamak**?

The development of controlled nuclear fusion has been a challenging task. Perhaps the most promising technique yet developed is called the tokamak, developed largely as the result of research by Russian physicist Lev Artsimovich (1909–1973) in the 1950s. The name "tokamak" is an acronym for "toroid camera with magnetic field." In a tokamak, nuclei are trapped in the middle of a magnetic field that has the shape of a "torus," a hollow, doughnut-shaped figure. The torus prevents particles from escaping from the field of reaction, turning them back onto themselves.

What is the **Nova Laser**?

The Nova Laser, of the Lawrence Livermore Laboratory, is the most powerful laser in the world. It directs ten laser beams to the center of its pellet chamber, where the laser creates a fusion reaction on the small fuel sample. Thus far the laser has been used for nuclear weapons research, and it is hoped that Nova will also help physicists make breakthroughs in the field of nuclear energy.

Since a major obstacle in obtaining fusion is the containment of plasma, has anyone been able to create **cold fusion**?

In March of 1989, two scientists by the names of Stanley Pons and Martin Fleischmann became overnight sensations when they claimed they had created cold fusion. Cold fusion would eliminate the problems of plasma containment, would save a great deal of money, and would, theoretically, provide the world with unlimited energy sources. Although their discovery sounded good, scientists were unable to reproduce the Pons and Fleishmann cold fusion. The fame and admiration that the two scientists received on their announcement quickly turned into embarrassment.

When will the world be able to **benefit from and use** fusion power?

Although the energy gained from fusion would be unlimited, the high cost of heating the atoms to a plasma state makes fusion economically unfeasible. However, many scientists feel that over the next forty to fifty years, we will have the ability to supply the majority of the world with fusion power. That's important, for those same scientists feel that in less than one hundred years we will have used up all the conventional energy sources in the world. Billions of dollars and a lot of time have gone into developing new methods for obtaining nuclear fusion. Eventually, science will find a way.

NUCLEAR WEAPONS

Who is considered to be **"The Father of the Atomic Bomb"**?

It was Albert Einstein, the famed German-American physicist who resided in the quiet town of Princeton, New Jersey, who has been called by many "The Father of the Atomic Bomb." Einstein developed the equation $E = mc^2$, which states that mass and energy are different manifestations of the same thing. This was the basic theory behind the nuclear reaction needed in the first bomb. It was also Albert Einstein who, along with other physicists, drafted a letter (the letter was actually written by Leo Szilard) to President Franklin D. Roosevelt, describing the bomb and encouraging a national effort for is development. Many historians and physicists feel that the reason Einstein participated in drafting the letter was to keep up with the Nazis, who were developing a bomb of their own.

The inquisitive, brilliant, and peace-loving Einstein was aware that people called him "the father of the atomic bomb" and was supposedly forever ashamed of that branding. In fact, some historians (and there is some controversy on this) say that Einstein, who did originally campaign for the nuclear bomb's development (although he never participated in its development), wrote to the President, pleading not to drop the bomb on Japan, but instead just threaten to use it to get the Japanese to surrender. His efforts, if the story is true, failed.

What was the **Manhattan Project**?

The U.S. government moved slowly on the letter from Einstein and other scientists during World War II that urged the development of an atomic bomb. It was not until mid-1942 that a program began in earnest; at that point, President Roosevelt authorized the creation of the Manhattan Project. The Manhattan Project was the name given the new Manhattan Engineering District created within the Army Corps of Engineers. Brigadier General Leslie R. Groves of the Corps was put in charge of the Project, and it was directed by physicist Robert Oppenheimer. This group of scientists (including Enrico Fermi) and engineers conducted high-level and highly secretive research and development on the design, construction, and detonation of an atomic bomb.

The ballistic casings of "Fat Man" (background) and "Little Boy," the atomic bombs dropped on Hiroshima and Nagasaki, Japan, in August 1945.

Where did the **first nuclear explosion** take place?

On July 16, 1945, the first nuclear explosive device was tested and detonated by the Manhattan Project in Alamogordo, New Mexico. It was a research device that was in no way, shape, or form suitable for delivery as a weapon. The scientists in the Manhattan Project referred to the explosion in Alamogordo as "the device"—it was really a test of a sub-critical mass plutonium implosion weapon, just like the bomb that would be dropped on Nagasaki. The first test of a weapon was really over Hiroshima. The Hiroshima bomb was a U-235 bomb; it was untested and it wasn't thought that it needed to be tested. The design was so simple that there was no doubt it would work. The design of the implosion bomb was much more difficult and warranted a test.

What was the **first nuclear bomb** dropped?

After the successful nuclear explosion in New Mexico, the United States government decided to end World War II with the Japanese by dropping the first nuclear weapon ever detonated over Hiroshima, Japan, on August 6, 1945. This atomic bomb, nicknamed "Fat Man," was so destructive because the Uranium-235 inside the bomb experienced a nuclear fission reaction. Detonated just above the city, the bomb created a great deal of damage through its initial blast, fireball, and radioactive radiation. In all, a third of the city was destroyed, leaving almost 130,000 dead and 180,000 homeless.

Three days later, the United States dropped a second bomb (known as "Little Boy") on the Japanese city of Nagasaki. Once again, a third of the city was destroyed, and 66,000 died.

Is there a **national organization or group** that watches over the **use of nuclear energy**?

Immediately following World War II, the U.S. Atomic Energy Commission was formed to regulate the use of nuclear energy. Although the Commission was abolished in the mid-'70s, the watchdog role was transferred to the Nuclear Regulatory Commission. The NRC, which is now a division of the Department of Energy, is in charge of all aspects of nuclear energy use and production. The international equivalent of the NRC is the International Atomic Energy Agency.

THE GRAND UNIFIED THEORY

What are the **four fundamental forces** that exist in the universe?

According to physicists, there are four fundamental forces that control how the universe works. The first and most obvious is the gravitational force. This is an attractive force that acts on all objects of mass. In order to truly feel the effects of gravity, an object must be in the vicinity of an extremely massive object such as the Earth, moon, or sun. The second force is actually a unification of two forces—the electric and the magnetic forces—combined by James Maxwell in 1864. He named this the electromagnetic force. This force is responsible for the transmission of light and the rest of the electromagnetic spectrum. The weak nuclear force, responsible for radioactive decay, is the third force. The weak force is basically a contact force, that is, two particles interact weakly only if they are in contact, or within very short range of each other. Finally, the strong nuclear force, the strongest of the four forces at extremely short distances, keeps the protons, neutrons, and other subatomic particles together in the nucleus.

What forces have been **unified**?

James Maxwell joined the electric and magnetic force in 1864, but it would be over 100 years until the electromagnetic and weak forces were

What is the Grand Unified Theory?

It is a widely held belief among many physicists that the fundamental forces of nature once existed as a single force from which the other forces evolved. This theory is called the Grand Unified Theory, or GUT. This comprehensive theory would explain how the forces of today have emerged into several separate forces.

proven to have been one force. Since 1973 and the unification of the electroweak force, physicists have not been able to unify the strong force with the electroweak force. Such unification would be a major step forward in understanding how the universe began.

What important **discovery** did Sheldon Glashow, Abdus Salam, and Steven Weinberg make in **unification theory**?

In 1961, Sheldon Glashow theorized that the electromagnetic force and the weak force could have been one unified force. Although he could not prove this experimentally, Glashow said that it could be proven through the discovery of the W and Z bosons, which are subatomic particles that mediate the weak nuclear force. Abdus Salam and Steven Weinberg were actually able to merge the electromagnetic and weak forces into the electroweak force, not through the discovery of the W and Z particles, but by using a concept called "symmetry," which can predict the behavior of what seems to be irregular behaviors of subatomic particles. In 1979, the three physicists received the Nobel prize in physics as a result of their work on symmetry, and for the unification and development of the electroweak force.

When were the **W and Z bosons** found?

Although Glashow, Weinberg, and Salam had received the Nobel prize for their theoretical solution to electroweak unification, it was not until

1983 that the scientific world would finally see the W and Z bosons. It was an Italian physicist named Carlo Rubbia who found the W and then Z boson at the CERN particle accelerator in Switzerland. For conclusive proof that the electroweak force existed, Carlo Rubbia and Simon van der Meer, the Dutch physicist who worked with Rubbia to find the W and Z, won the 1984 Nobel prize in physics.

Who **developed** the Grand Unified Theory?

The main player in the development of the Grand Unified Theory was Albert Einstein. He started working on calculations and theories that might one day unify the forces of gravity and electromagnetism (the only forces known to Einstein at that time). Einstein was not able to solve the problem of unification, for during his lifetime two additional forces—the strong and the weak—were discovered, making the challenge of unification much more complex.

THE THEORY OF EVERYTHING

What is the difference between the **Grand Unified Theory** and the **Theory of Everything**?

Both the Grand Unified Theory and the Theory of Everything attempt to explain how the universe behaved right after the big bang (see later in this chapter for discussion of the big bang). The Grand Unified Theory explains the origins of the electromagnetic, weak, and strong forces. Scientists can realistically unify these forces, for they all work on the principles of quantum mechanics. The final force, gravity, has not yet been proven to be governed by quantum mechanics, and will therefore be extremely difficult to merge or unify with the other forces that do rely on quantum mechanics. If physicists can one day fuse gravitational force with the Grand Unified forces, it will be known as the "Theory of Everything." This, of course, is the ultimate goal for many physicists, and if achieved, would explain a lot about what happened in the first few moments of the universe.

What are the **four dimensions** of our world?

The world in which we live is generally thought to have four dimensions. The first three describe space, and they are referred to as x (length), y (width), and z (height). For example, a single dimension would be like a string; you would only be allowed to move forward and backward in that one-dimensional world. To broaden the existence to a two-dimensional world would be to live as a drawing on a piece of paper; on that two-dimensional world, you would be able to move forward and backward as well as side to side. Then if a third dimension were added, you would be able to leave the paper to move up and down. Finally, a fourth dimension would not measure your position in the world through the x, y, and z coordinates, but instead it would measure time, which is what allows us to remember the past and move into the future. This is why we are considered to live in a four-dimensional world, called space-time.

What is **string theory**?

String theory, the thought that all particles could actually be made up of strings, has become one of the most exciting areas of physics in recent

Is there a dimension beyond that which is known to man?

Mathematician Theodor Kaluza originally proposed a fifth dimension in the early 1920s. Einstein had a difficult time accepting the idea of a fifth, tightly wrapped dimension, and did not use it in his theories on unification. Some physicists and mathematicians feel that there are more than four or five dimensions, and that there could be as many as ten or eleven dimensions. To imagine a dimension beyond the four we experience would be almost impossible, for we are not aware of experiencing such a thing.

years. The theory that particles could actually exist as multidimensional strings has been around since the 1960s, but has only been taken seriously over the last decade or so. If multidimensional strings actually existed, then according to some physicists and mathematicians, the Theory of Everything would be easier to achieve. However, for many non-physicists, the idea that there could be more dimensions than the four of which we are now aware is difficult to comprehend.

THE EXPANDING UNIVERSE

What is **cosmology**?

Cosmology, a branch of physics and astronomy, focuses its efforts on understanding and documenting the history and future of the universe. Since the earliest days of history, humankind has struggled to understand the world in which we live. Over the last 2,000 years, that "world" has expanded to include the starry heavens as well as the Earth itself. We now use the term "universe" to mean all that was, is, and will be available to observation and measurement. In the twentieth century, the scale of the universe has been pushed beyond our local galaxy of about 100 billion stars to include similar galaxies flung across space, to the most distant reaches observable with the largest telescopes. Cosmology is an attempt to describe the large-scale structure and order of the universe. To that end, one does not expect cosmology to deal with the detailed structures of planets, stars, or even galaxies. Rather it attempts to describe the structure of the universe on the largest of scales and to determine its past and future.

Who theorized that the universe is actually **expanding**?

Edwin Hubble, in the early twentieth century, studied the size, distance, and motion of different stars and galaxies in the universe. As a result of his and another astronomers' observations, he theorized that the universe was still expanding. To back up such a major and controversial

What was "the biggest blunder" of Einstein's life?

Albert Einstein is reputed to have made this self-deprecating remark after making one of the few scientific mistakes of his life. Einstein, as well as most physicists of the early twentieth century, incorrectly believed that the universe was a static body (not moving). Einstein usually relied on the beauty of mathematical equations to be proof enough that they hold true. But in his equations dealing with the universe, he introduced an anti-gravity constant, known as the cosmological constant, that never should have been included. If Einstein had relied on the pure equations and not created a constant, he would have been the first to predict an expanding universe and would not have made such a big "blunder."

statement, he proved that most of the galaxies and stars he had observed experienced a red shift in their spectral lines. (A red shift, the result of the Doppler effect, means that an object is moving away from the Earth, while a blue shift means it is moving toward the Earth.) This was an experimental triumph, and completely unexpected at the time. It was the crucial piece of evidence that finally led to the development of the big bang theory. Although Einstein could have made the same prediction through his calculations, he unfortunately made a mistake by introducing a false constant (see above question).

The Big Bang

What is the **big bang**?

The big bang is the model that scientists use to describe the creation of the universe. It states that the universe was created in a violent event that occurred approximately 10 to 20 billion years ago. In that event, the lightest elements were formed, which provided the building blocks for all of the matter that exists in the universe today. A consequence of the

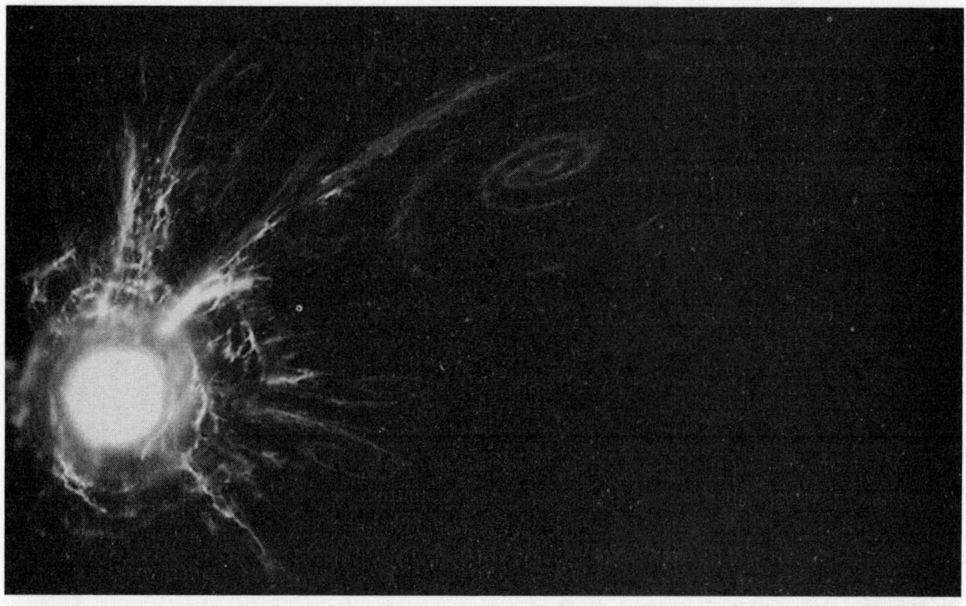

An artist's impression of galaxies being formed in the aftermath of the big bang. The spiral clouds of gas have already started condensing into the shapes of future galaxies.

big bang is that we live in an expanding universe, the ultimate fate of which cannot be predicted from the information we have at this time.

Who **first suggested** that the universe began with a big bang?

In 1927, a Belgian priest named Georges Lemaitre claimed the universe had a definite beginning. He felt it began when a heavy and compressed atom exploded in a massive bang. Lemaitre published this idea in a Belgian journal, but it was not taken seriously until Hubble's work on the expanding universe. Although Lemaitre's ideas on the big bang do not completely agree with what is believed today, it was a big first step for theories dealing with the beginning of the universe.

How did **cosmic radiation** help prove that the big bang actually occurred?

In 1964, two scientists from Bell Laboratories in New Jersey heard, by accident, the hissing sound of cosmic radiation coming from all direc-

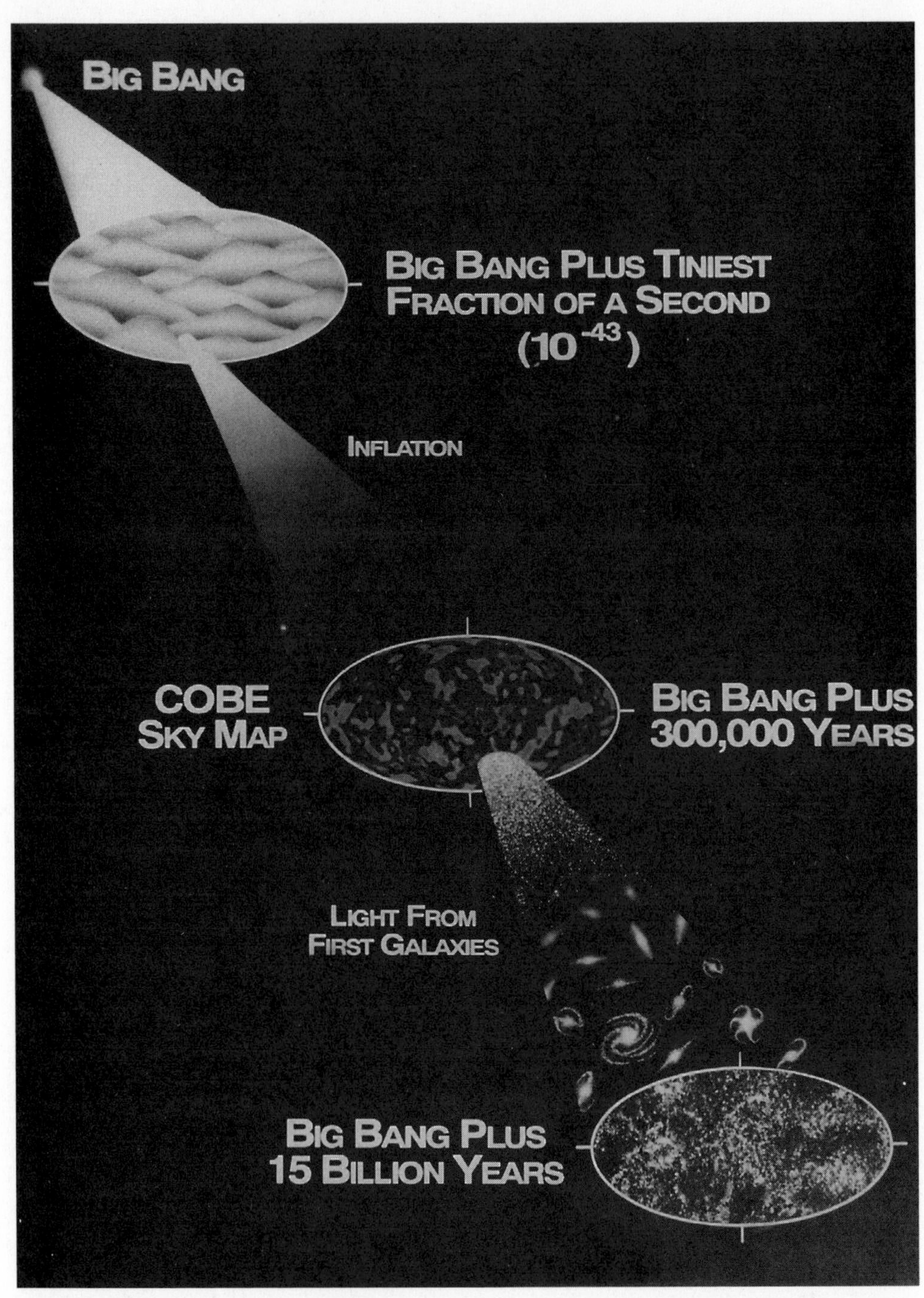

Diagram depicting crucial periods in the development of the universe. The top oval represents the big bang, the next oval shows a sky map measuring background radiation, and the final oval shows galaxies and stars beginning to form.

tions in the universe. It was the discovery of such background noise that proved to be such a major breakthrough for the big bang theory. Until then, some scientists felt that there was no beginning or end to the universe, but that the universe had always existed—and would forever remain—in a relaxed state.

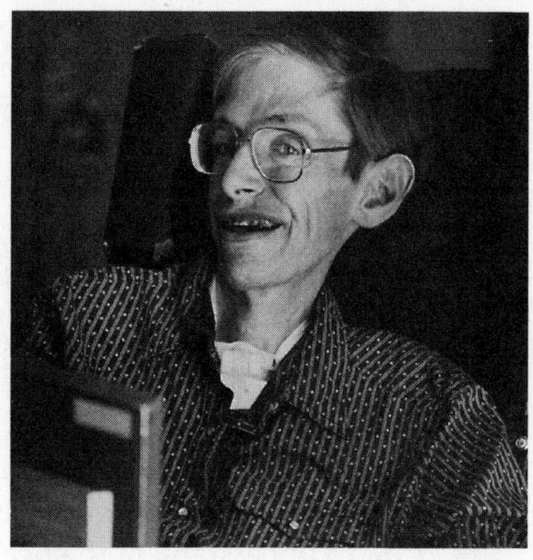
Stephen Hawking.

Who is **Stephen Hawking**?

Stephen Hawking is probably one of the most well-respected and famous scientists of our time. His battle with amyotrophic lateral sclerosis, a debilitating disease that affects the nervous system, has confined him to a wheelchair and voice synthesizer, but has not slowed down his ability or desire to be the most influential physicist of the late twentieth century.

Hawking has conducted most of his studies in the field of cosmology. Specifically, he has been instrumental in proving the existence of black holes, and bridging the gap between Einstein's Theory of Relativity and the big bang theory. Hawking has written popular books on our universe, including *A Brief History of Time.*

What do physicists believe will **eventually happen** to the universe?

There are three main theories concerning the fate of the universe. The first is called the open universe. It says that the universe is expanding and will continue expanding at this rate forever. The second theory, called the closed universe, is quite popular and says that the universe is expanding, but it will eventually come to a stop and collapse. Some even feel that after a collapse, another big bang will take place. The final theory involves a marginally open universe. Although this theory does not hold that the universe will expand forever, it also says that the universe will not collapse either. In fact, it is believed that the expansion rate of the universe will slow down to such a crawl as to virtually cease its expansion.

NEUTRINOS

What is a **neutrino**?

Neutrinos are tiny, elusive subatomic particles that were produced from the big bang and are emitted by the sun and every other star throughout the universe. Until recently not much was known about the neutrino except for the fact that there are a lot of them— they are approximately a billion times more prevalent throughout the universe than common particles that make up atoms.

Neutrinos were first suggested to exist in 1930, but were not found experimentally until 1956. They are deemed the strangest and most difficult to understand of the twelve fundamental subatomic particles that make up all matter. As with quarks, there are flavors of neutrinos that are used to characterize the different types. These flavors are the electron, muon, and tau neutrino. The reason why neutrinos are so difficult to understand is that they are difficult to observe. In fact, physicists feel that tens of thousands of neutrinos pass through any person's body every second and that in every electron there could be as many as 50 billion neutrinos! Since they are so small, and carry no electrical charge, physicist have had a very difficult time studying them. Until now.

What important breakthrough has been made in the observation of neutrinos?

Theoretical physicists have argued about the possibility of the neutrino having mass, and the effects on the universe if it did. Since there are so many neutrinos throughout the universe, and knowing the particle's mass (if it had any) would be important, the problem seemed to be a big stumbling block. Until June 1998, that is.

According to physicists working at a Japanese facility located 2,000 feet underground, if the neutrino can oscillate or change flavors or types, then according to quantum physics, the neutrino has mass.

Where was the neutrino found to have had mass?

It was in Super-Kamiokande, a 12.5 million–gallon stainless steel–lined underground super-pure water tank buried 2,000 feet under a Japanese mountain, that the neutrino's mass was discovered. The neutrinos would be observed about once every hour and a half, when neutrino interactions would produce faint flashes of light, picked up by 13,000 light sensors. The reason the tank was underground was mostly to eliminate other cosmic rays and other particles that would interfere with the detection of neutrinos. The collaborative, $100-million effort was sponsored by the University of Tokyo's Institute for Cosmic Ray Research, which included eight Japanese and six American universities.

What does the neutrino's mass **mean** to the **universe**?

The discovery of the neutrino's mass sent physicists throughout the world back to their drawing boards, attempting answering new questions about the universe. Since all the stars, galaxies, and planets throughout the universe account for only 10% of the universe's mass, how much of the remaining 90% is the neutrino responsible for? And what will be the impact on the universe? Many argue that since neutrinos have mass, they are also attracted toward each other through gravitational force. Perhaps in a few billion years, the universe's expansion will cease, and the universe will contract due to the gravitational attraction that all the neutrinos have for each other. These are only theories, and in the years to come, physicists should have more answers, more theories, and more questions.

RELATIVITY

What is **Relativity**?

Albert Einstein's Theories of Relativity consist of two major portions: the Special Theory of Relativity and the General Theory of Relativity. Special Relativity deals with phenomena that become noticeable when traveling near the speed of light, and with reference frames that are moving at a constant velocity (inertial reference frames). General Relativity deals with reference frames that are accelerating (noninertial reference frames), and with phenomena that occur in strong gravitational fields. General Relativity also uses the curvature of space to explain gravity.

Special Relativity

What is the **Special Theory** of Relativity?

In the Special Theory, his first theory of Relativity, Einstein declared that speed is never constant, for motion is always relative to one's perspective. However, Einstein did declare that light, no matter how fast an observer is travelling, always travels at 3×10^8 m/s (meters per second) in a vacuum. In this theory, Einstein also describes how motion can affect time, and the relationship between mass and energy is explained in the equation $E = mc^2$.

People before Einstein had tried to calculate fields for bodies moving at high velocities, but Einstein alone was able to do this correctly by making a number of what was then radical assumptions, one of which was that the speed of light were an upper limit to velocities, and the speed of light was the same for all observers. This theory had no experimental evidence at the time, and was considered near to heresy by many in the field. What makes Special Relativity "special" is that Einstein only considered cases in which objects had constant velocities; accelerations were covered by the succeeding "General Relativity."

What is **time dilation**?

Time dilation plays a major role in Einstein's Theory of Relativity, which states that if someone were able to travel at the speed of light, time would stop. Since no one can actually travel that fast, the rule has been explained as time dilation: the faster one travels, the slower time moves. This means that if someone is moving quickly through space, time compensates for that motion—for, according to the Special Theory of Relativity, the laws of physics are the same no matter what a person's speed.

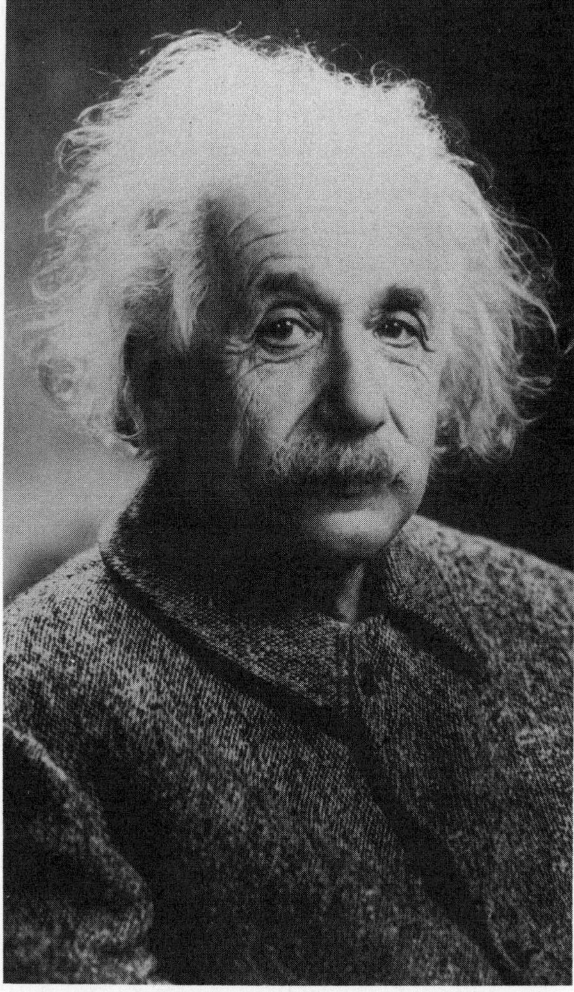

Albert Einstein.

Jets flying at high velocities for extended periods of time have proved time dilation in the Special Theory. Using extremely accurate atomic clocks, physicists found that time actually moved slower on a fast-moving airplane. The slowing of time does not become humanly noticeable until speeds close to the speed of light are obtained. For example, according to Einstein's formulas, if someone were to travel one-tenth the speed of light, not nearly possible today, time would slow down by only 0.5%. In fact, the fastest spaceships would slow down time by only two millionths of one percent. These results are not humanly recognizable and will never be until we can achieve speeds close to the speed of light.

What is the twin paradox?

The twin paradox is an example that attempts to illustrate the effects of time dilation. Einstein claimed that if one twin boarded a spaceship and traveled close to the speed of light, time would slow down for that twin, while for the twin on Earth, time would be normal. Both twins would feel like time passed at a normal rate, but upon the first twin's return to Earth, he may have only aged a few months while the twin left on Earth might be old and gray.

GENERAL RELATIVITY

What is the **General Theory** of Relativity?

Many were impressed with Einstein's ideas of Special Relativity, but even more were blown away by the theories of General Relativity. In fact, Einstein's General Theory of Relativity was so "deep" that perhaps only a few dozen people in the world at that time truly understood what Einstein's theory was about.

Whereas the Special Theory dealt with the relationship between motion and time, the General Theory dealt with the connection between mass and motion. According to Einstein, the faster a body accelerates, the slower time moves for that body, and the greater its mass.

Another aspect of the General Theory says that massive objects (which have large gravitational forces) actually deform and warp space. Einstein predicted that massive objects that warp space could actually bend light. His theory was proved after a solar eclipse in 1919. At night, the location of a star is easily identified; however, during the day—when the sun is right next to that particular star—the star's position will appear to have changed position. Einstein had to wait for a solar eclipse because otherwise, the blinding light of the sun would make it impossible to see the star. Einstein, with his reputation on the line during this observation, won huge fame as a result of the discovery.

EINSTEIN RINGS

What is an **Einstein Ring**?

Albert Einstein's General Theory of Relativity says that a massive object such as a star or galaxy warps the space around it, causing light to bend as it approaches that massive object. Einstein's theory was first proved during a solar eclipse in 1919 when the position of a star seemed to have shifted when the sun was in close alignment with that star.

Recently, one of Einstein's lesser-known theories on warped space was proven to exist. It is called an "Einstein Ring," and is also a result of space warping the path of light. It says that if a massive object such as a galaxy, in this case called a lens galaxy, is in direct alignment with another massive stellar object behind it, then a ring of light (the image of the distant object) should appear around the lens galaxy. Without knowledge of Einstein's theories, one might think that the distant object would be obscured by the galaxy in front of it, but according to Einstein, the light from the distant object uniformly bends around the lens galaxy, producing an out-of-focus image completely around the galaxy.

Has anyone ever **seen** an Einstein Ring?

A group of astronomers claimed they had discovered what appeared to be a partial Einstein Ring. The partial ring meant that the image of the distant object was seen as an arc around the lens galaxy, and not a complete circle. The astronomers, who made this discovery in 1979, claimed that it was partial because the distant object, the lens galaxy, and the Earth were not in perfect alignment. Since 1979, approximately twenty or so partial Einstein Rings have been found.

However, in March of 1998, astronomers made a fascinating discovery. With the help of radio telescopes throughout Great Britain and the Hubble Space Telescope, British and American astronomers finally observed the first complete Einstein Ring. The event proved that Einstein's theory, proposed eighty years earlier, was indeed correct.

BLACK HOLES

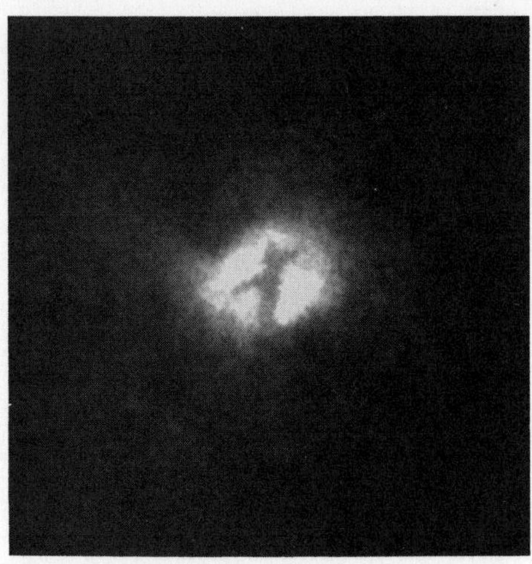

An image of the core of the Whirlpool galaxy that shows an immense ring of dust and gas (the X-shaped dark lines at the center) that is thought to surround and hide a giant black hole, one million times the mass of the sun, in the center of the galaxy.

What is a **black hole** and what does it do to light?

Stephen Hawking has made wonderful discoveries and breakthroughs with respect to black holes, but his theories are derived from the theories of Einstein. Black holes are incredibly massive objects with huge gravitational fields that warp the space around it. If a star, planet, and even light get too close to a black hole, the severely warped space around the black hole will cause the matter or light to spiral down into the hole, just as water does as it moves toward the drain of a sink. By combining Einstein's Theory of Relativity with the ideas of quantum theory—two theories previously thought to contradict each other—Hawking showed how the escape velocity of a black hole (that is, the velocity an object must attain in order to "escape" the pulling force of a massive object) exceeds the speed of light—and that since nothing can exceed the speed of light, nothing, including light, can escape the black hole.

Further Reading

Robert Kemp Adair. *The Physics of Baseball.* HarperCollins, 1995 (second edition). $12.00 (softcover). ISBN: 0060950471.

Jonathan Allday. *Quarks, Leptons, and the Big Bang.* Iop Publications/Institute of Physics, 1997. $29.50 (softcover). ISBN: 0750304626.

Isaac Asimov. *Atom: A Journey across the Subatomic Cosmos.* Penguin USA, 1992 (reprint edition). $15.95 (softcover). ISBN: 0452268346.

Leonid V. Azaroff. *Physics over Easy: Breakfasts with Beth and Physics.* World Scientific Pub, 1996. $18.00. ISBN: 9810223676.

J. D. Bernal. *A History of Classical Physics.* Barnes & Noble Books, 1997 (reprint edition). $12.98. ISBN: 0760706018.

Jeremy Bernstein. *Albert Einstein and the Frontiers of Physics.* Oxford University Press Children's Books, 1997. $12.95 (softcover). ISBN: 0195120299.

Jeremy Bernstein. *Cranks, Quarks, and the Cosmos.* Basic Books, 1994. $12.00 (softcover). ISBN: 0465014496.

Louis A. Bloomfield. *How Things Work: The Physics of Everyday Life.* John Wiley & Sons, 1996. $58.85 (softcover). ISBN: 0471594733.

Craig F. Bohren; foreword by Jearl Walker. *Clouds in a Glass of Beer: Simple Experiments in Atmospheric Physics.* John Wiley & Sons, 1987. $17.95 (softcover). ISBN: 0471624829.

Denis Brian. *Einstein: A Life.* John Wiley & Sons, 1997. $19.95 (softcover). ISBN: 0471193623.

Carnegie Library of Pittsburgh, Science and Technology Department. *The Handy Science Answer Book.* Visible Ink Press, 1997 (second edition). $16.95 (softcover). ISBN: 0787610135.

David C. Cassidy. *Uncertainty: The Life and Science of Werner Heisenberg.* W. H. Freeman & Company, 1993 (reprint edition). $21.95 (softcover). ISBN: 0716725037.

D. Hatcher Childress. *The Anti-Gravity Handbook.* Adventures Unlimited Press, 1993 (revised edition). $14.95 (softcover). ISBN: 0932813208.

Robert Ehrlich, Jearl Walker. *Turning the World Inside Out and 174 Other Simple Physics Demonstrations.* Princeton University Press, 1990. $16.95 (softcover). ISBN: 0691023956.

Robert Ehrlich. *Why Toast Lands Jelly-Side Down: Zen and the Art of Physics Demonstrations.* Princeton University Press, 1997. $14.95 (softcover). ISBN: 0691028877.

Albert Einstein. *The Meaning of Relativity.* Fine Communications, 1997 (fifth edition). ISBN: 1567311369.

Albert Einstein. *The Principle of Relativity.* Dover Publications, 1924. $7.95 (softcover). ISBN: 0486600815.

Albert Einstein. *Relativity: The Special and the General Theory.* Crown Publications, 1995 (reprint edition). $7.00 (softcover). ISBN: 0517884410.

Albert Einstein. *The World As I See It.* Citadel Press, 1993 (reprint edition). $9.95 (softcover). ISBN: 080650711X.

Phillis Engelbert, Diane L. Dupuis. *The Handy Space Answer Book.* Visible Ink Press, 1998. $17.95 (softcover). ISBN: 1578590175.

Barry Evans. *Everyday Wonders: Encounters with the Astonishing World around Us.* NTC/Contemporary Publishing, 1993. $16.95 (softcover). ISBN: 0809237989.

David Filkin; foreword by Stephen Hawking. *Stephen Hawking's Universe: The Cosmos Explained.* Basic Books, 1997. $30.00. ISBN: 0465081991.

Ritchie Fliegler, Jon F. Eiche. *Amps!: The Other Half of Rock 'n' Roll.* Hal Leonard Publishing, 1993. $24.95 (softcover). ISBN: 0793524113.

Albrecht Folsing. *Albert Einstein: A Biography.* Penguin USA, 1998. $18.95 (softcover). ISBN: 0140237194.

Harald Fritzsch, Karin Heusch. *An Equation That Changed the World: Newton, Einstein, and the Theory of Relativity.* University of Chicago Press, 1994. $29.95. ISBN: 0226265579.

Murray Gell-Mann. *The Quark and the Jaguar: Adventures in the Simple and the Complex.* W. H. Freeman & Company, 1995 (reprint edition). $15.95 (softcover). ISBN: 0716727250.

Don Glass, Stephen Fentress. *Why You Can Never Get to the End of the Rainbow and Other Moments of Science.* Indiana University Press, 1993. $10.95 (softcover). ISBN: 0253207800.

Donald Goldsmith. *Einstein's Greatest Blunder?: The Cosmological Constant and Other Fudge Factors in the Physics of the Universe.* Harvard University Press, 1997 (reprint edition). $14.95 (softcover). ISBN: 0674242424.

Larry Gonick, Art Huffman. *The Cartoon Guide to Physics.* HarperPerennial, 1992. $15.00 (softcover). ISBN: 0062731009.

John Gribbin. *In Search of Schrodinger's Cat: Quantum Physics and Reality.* Bantam Doubleday, 1985 (reprint edition). $13.95 (softcover). ISBN: 0553342533.

Michael Guillen. *Five Equations That Changed the World: The Power and Poetry of Mathematics.* Hyperion, 1996. $13.95 (softcover). ISBN: 0786881879.

Stephen Hawking. *Black Holes and Baby Universes and Other Essays.* Bantam Books, 1994 (reprint edition). $14.95 (softcover). ISBN: 0553374117.

Stephen Hawking. *A Brief History of Time: From the Big Bang to Black Holes.* Bantam Books, 1990 (reprint edition). $14.95 (softcover). ISBN: 0553346148.

Werner Heisenberg. *Encounters with Einstein and Other Essays on People, Places, and Particles.* Princeton University Press, 1989 (reprint edition). $10.95 (softcover). ISBN: 0691024332.

Werner Heisenberg. *Physical Principles of the Quantum Theory.* Dover Publications, 1930. $7.95 (softcover). ISBN: 0486601137.

Nick Herbert. *Faster Than Light: Superluminal Loopholes in Physics.* New American, 1995 (reprint edition). $12.95 (softcover). ISBN: 0452263174.

Nick Herbert. *Quantum Reality: Beyond the New Physics.* Anchor, 1987 (reprint edition). $11.95 (softcover). ISBN: 0385235690.

Tony Hey, Patrick Walters. *Einstein's Mirror.* Cambridge University Press, 1997. $27.95 (softcover). ISBN: 0521435323.

Roger S. Jones. *Physics for the Rest of Us: Ten Basic Ideas of 20th Century Physics That Everyone Should Know . . . and How They Shaped Our Culture and Consciousness.* NTC/Contemporary, 1993. $18.95 (softcover). ISBN: 0809237164.

Lawrence M. Krauss. *Beyond Star Trek: Physics from Alien Invasions to the End of Time.* Basic Books, 1997. $21.00. ISBN: 046500637X.

Lawrence M. Krauss; foreword by Stephen Hawking. *The Physics of Star Trek.* Harper-Perennial, 1995. $12.00 (softcover). ISBN: 0060977108.

Sven Kullander, Borje Larsson. *Out of Sight: From Quarks to Living Cells.* Cambridge University Press, 1994. $32.95. ISBN: 0521350441.

Leon Lederman, Dick Teresi. *The God Particle: If the Universe Is the Answer, What Is the Question?* Delta, 1994 (reprint edition). $12.95 (softcover). ISBN: 0385312113.

Robert L. Lehrman. *Physics the Easy Way.* Barrons, 1998 (third edition). $12.95 (softcover). ISBN: 0764102362.

Robert P. Libbon. *The Ultimate Einstein.* Pocket Books, 1997. $30.00 (includes CD-ROM). ISBN: 0671011715.

David Lindley. *The End of Physics: The Myth of a Unified Theory.* Basic Books, 1994 (reprint edition). $14.00 (softcover). ISBN: 0465019765.

David Lindley. *Where Does the Weirdness Go? Why Quantum Mechanics Is Strange, but Not As Strange As You Think.* HarperCollins, 1997. $13.00 (softcover). ISBN: 0465067867.

Walter A. Lyons. *The Handy Weather Answer Book.* Visible Ink Press, 1997. $16.95 (softcover). ISBN: 0787610348.

Robert H. March. *Physics for Poets.* McGraw-Hill, 1995 (fourth edition). $35.65. ISBN: 0070402485.

Harold J. Morowitz. *The Thermodynamics of Pizza.* Rutgers University Press, 1992 (reprint edition). $14.95 (softcover). ISBN: 0813517745.

Lloyd Motz, Jefferson Hane Weaver. *The Story of Physics.* Plenum Press, 1989. $24.50. ISBN: 0306430762.

Robert Osserman. *Poetry of the Universe.* Anchor Books, 1996. $10.95 (softcover). ISBN: 0385474296.

Abraham Pais. *Niels Bohr's Times: In Physics, Philosophy, and Polity.* Oxford University Press, 1993 (reprint edition). $17.95 (softcover). ISBN: 0198520484.

Sybil P. Parker. *The McGraw-Hill Dictionary of Physics.* McGraw-Hill, 1996 (second edition). $16.95 (softcover). ISBN: 0070524297.

Thomas Powers. *Heisenberg's War: The Secret History of the German Bomb.* Little Brown & Company, 1994 (reprint edition). $16.95 (softcover). ISBN: 0316716235.

Martin J. Rees; foreword by Stephen Hawking. *Before the Beginning: Our Universe and Others.* Perseus Press, 1997. $25.00. ISBN: 0201151421.

B. K. Ridley. *Time, Space, and Things.* Cambridge University Press, 1995 (third edition). $9.95 (softcover). ISBN: 0521484863.

Tony Rothman. *Instant Physics: From Aristotle to Einstein, and Beyond.* Fawcett Books, 1995. $10.00 (softcover). ISBN: 0449906973.

Joseph Schwartz, Michael McGuinness. *Einstein for Beginners.* Pantheon Books, 1990 (reprint edition). $11.00 (softcover). ISBN: 0679725105.

Cindy Schwarz; introduction by Sheldon Glashow. *A Tour of the Subatomic Zoo: A Guide to Particle Physics.* American Institute of Physics, 1996 (second edition). $29.95 (softcover). ISBN: 1563966174.

Curt Suplee. *Everyday Science Explained.* National Geographic Society, 1996. $35.00. ISBN: 0792234103.

Henk Tennekes. *The Simple Science of Flight: From Insects to Jumbo Jets.* MIT Press, 1997. $12.50 (softcover). ISBN: 0262700654.

Kip S. Thorne; foreword by Stephen Hawking. *Black Holes and Time Warps: Einstein's Outrageous Legacy.* W.W. Norton & Company, 1995 (reprint edition). $15.95 (softcover). ISBN: 0393312763.

Robert Trostli. *Physics Is Fun: A Sourcebook for Teachers.* Octavo Editions, 1997. $30.00 (softcover). ISBN: 0964276046.

Steven Weinberg. *Dreams of a Final Theory.* Vintage Books, 1994 (reprint edition). $13.00 (softcover). ISBN: 0679744088.

Steven Weinberg. *The First Three Minutes: A Modern View of the Origin of the Universe.* Basic Books, 1993 (second edition). $14.50 (softcover). ISBN: 0465024378.

Index

Note: Page numbers in bold refer to illustrations.

B

C

I

J

K

L

M

N

Q

R

S

U

V

W

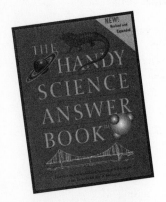

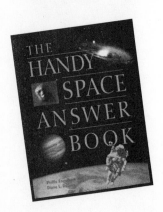